Laboratory Exercises in Human Physiology

Laboratory Exercises
in
Human Physiology

Paul M. Spannbauer

Hudson Valley Community College

1817

HARPER & ROW, PUBLISHERS, New York
Cambridge, Philadelphia, San Francisco,
London, Mexico City, São Paulo, Sydney

Laboratory Exercises in Human Physiology
Copyright © 1983 by Harper & Row, Publishers, Inc.

ISBN 0-06-046372-4

CONTENTS

vi

Audience

   Laboratory Exercises in Human Physiology is designed for use in a one-semester introductory course in physiology. The experiments are geared to students in biological, medical, and allied health programs. Among the students who would benefit from this lab manual are those in nursing, physical therapy, inhalation therapy, mortuary science, physician's assistant, and other allied health sciences. In addition, the lab manual is appropriate for students in the biological sciences, science technology, liberal arts, physical education, premedical, predental, and prechiropractic programs.

Objectives

   The objectives of this laboratory manual are:

1. To provide laboratory experiences that demonstrate basic physiological principles.
2. To provide students with an opportunity to become familiar with various types of laboratory instrumentation used in physiology.
3. To provide an opportunity for students to work together to accomplish a common goal.
4. To provide students with a feel for the type and quality of data gathered in physiological research.

Special Features

   The lab manual contains the following special features:

1. Behavioral Objectives. Each experiment begins with a list of behavioral objectives. Each objective describes a concept or skill which students should understand or be able to perform after doing the experiment.
2. Materials. Each experiment contains a list of materials needed. These lists should act as guides for preparation of the laboratory by instructors.
3. Introduction. Each experiment has an introduction which provides background information needed for a thorough understanding of the experimental procedure. Examples of typical data and sample problems are included where appropriate.
4. Experimental Procedure. Each experiment contains a detailed, step-by-step set of instructions for performing the experiment. These procedures have been classroom tested and have been found to work without technical difficulty. The procedures have been designed to take approximately 2 hours to accomplish.

Courses designed for 3-hour laboratory sessions could use the experiments for the first 2 hours and spend the last hour analyzing the data and completing the laboratory data sheets.

5. Data sheets. Each experiment has a data sheet associated with it. These data sheets are designed to provide all of the space required for responding to questions, graphing, and summarizing data. This should facilitate the completion of lab reports by students and the grading of lab reports by instructors.

6. Appendix. The appendix, *Experimental Solutions,* is designed as an aid to the instructor in the preparation of some experimental solutions.

Acknowledgments

As is usually the case, numerous people and events have contributed to the publication of this laboratory manual. I would like to thank Claudia Wilson and Mary Matson of Harper & Row, Publishers, for giving me the encouragement to prepare the manuscript for publication. I am grateful to the Literary Executor of the late Sir Ronald A. Fisher, F.R.S., to Dr. Frank Yates, F.R.S., and to Longman Group Ltd., London, for permission to reprint part of Table III—t distribution from their book *Statistical Tables for Biological, Agricultural, and Medical Research* (6th Edition, 1974).

I would like to thank Emily Kirby and Stephen Hyatt of Hudson Valley Community College for valuable suggestions on bookwriting in general. Finally, I would like to express my appreciation to my physician's assistant students, past and present, for constructive criticism of the experimental procedures.

Paul M. Spannbauer

EXPERIMENT 1

INSTRUMENTATION AND STATISTICS

Behavioral Objectives

The student should be able to:

1. Define a physiological variable.
2. List the components of an instrument.
3. List and define the characteristics of instruments.
4. Define some commonly used statistical terms.
5. Become familiar with the operation of the physiograph or kymograph and the stimulator.
6. Perform a *t* test of significance on sample data.

Materials

Physiograph or kymograph
Stimulator
Appropriate electrical cables

Introduction

1.1. Instrumentation and Physiological Variables. An *instrument* is a tool or implement that is used for a particular purpose. A wide variety of instruments are used in physiology. Some of these instruments are quite simple; others are quite complex (Table 1.1).

An important function of physiological instrumentation is the measurement of *physiological variables*. A variable is any quantity whose value changes with time, for example, blood pressure, heart rate, and respiration rate. In order to be useful, a physiological variable must be measurable. Some variables, for example, body weight, vary at a very slow rate, whereas other variables, for example, the electroencephalogram, vary rapidly. In some cases, it is necessary to measure the variable over a period of time and average the results before useful information can be obtained. At times, a variable can be measured only in response to the application of a stimulus. In such cases, the response to the stimulus is used to infer information about the variable.

1.2. Transducers, Conditioners, and Displays. All instruments can be divided into three basic components: a transducer, a signal conditioner, and a display. In addition, some instruments include a stimulator.

1

Table 1.1 Systemic Physiological Instrumentation

SYSTEM	ORGAN	QUANTITY	INSTRUMENT	MEASUREMENT
Cardiac	Heart	Pulse rate waveform	Sphygmomanometer Arrhythmia monitor Defibrillator Stethoscope	ECG Recorder Resuscitator O_2 Applicator Rate meter
Pulmonary	Lungs	Breath rate Tidal volume Thoracic resistance	Spirometer Respirator Nebulizer	Suction Pumps Fluoroscope
Circulatory	Blood	Pressure Flow rate Oxygen content CO_2 content Acidity Cell count	Pressure Flow } Meter Temperature pH meter Cell counters	Coagulometer Hemoglobinometer Chromatograph Electrophoresis
Urinary	Kidneys Bladder	Urine composition Electrolyte balance	Cell analyzers Densitometer Acid-base Balance pH Measurement	Osmometer
Nervous	Brain Spine Skin	Brain waves Reflexes PSGR	Encephalograph Echoencephalograph Angiometer	Psychogalvanometer
Digestive-gastro	Stomach Liver Intestines	Acidity Enzymes Fecal comp.	Endoscope Bilirubinometer Sigmoidoscope	Fluoroscopy
Muscular-skeletal	Torso Limbs Extremities	Flexure Stiffness Reflexes	Electromyograph Reflexometer X-rays	
Sensory	Eye Ear, smell Taste	Sound pressure Optical	Perimeter Tonometer Ophthalmoscope	Audiometer Retinoscope Salivivometer

Table 1.1 (Continued)

SYSTEM	ORGAN	QUANTITY	INSTRUMENT	MEASUREMENT
Endocrine	Pituitary Lymph Pancreas	Enzyme composition	Cell analyzers Densitometer Osmometer	
Reproductive	Gonads Genitals Ovaries	Hormone composition Acidity Fetal movement Labor action	Cardiotachograph Echoencephalograph	
Integumentary	Skin Orifices Hair	Touch Acidity Weight Texture	Plethysmoscope Thermograph Reflexometer Depilator	

Source: From Thomas, *HANDBOOK OF BIOMEDICAL INSTRUMENTATION AND MEASUREMENT*, 1974, p. 25. Reprinted with permission of Reston Publishing Co., a Prentice-Hall Co., 11480 Sunset Hills Road, Reston, VA 22090.

3

The *transducer* is the part of the instrument that senses the physiological variable to be measured. The variable must be converted to an electrical signal—a voltage that changes its value in exactly the same way as the physiological variable. In most instruments, the electrical signal produced by the transducer must be modified before it is suitable for use. This modification is called *signal conditioning* and usually involves amplification—making a small signal larger. The signal-conditioning equipment may consist of a simple amplifier or may be as sophisticated as a computer. It is usually housed within the cabinet of the instrument, and controls that alter the operation of this equipment are provided on the front panel of the instrument.

The results of a measurement must be displayed so that they can be perceived by one or more human senses. The part of an instrument that converts the output of the signal conditioner into perceivable form is the *display*. Typical displays include indicator lights, buzzers, meters, ink recorders, and visual monitors. At times, more than one form of display is used in an instrument. For example, a heart monitor may make use of a visual display for the electrocardiogram, a digital meter for the heart rate, and a buzzer for a danger signal.

Information gathered by several transducers may be displayed side by side on a single visual monitor or on a single piece of recording paper. The measurement from each transducer is called one *channel* of information. Equipment may be used to record this information for future reference or analysis.

1.3. Instrumentation Terminology.

In order to properly make use of instruments, it is important to understand the terms that describe the limitations of instruments. The following terms are used to describe various instruments:

Range—the set of values over which an instrument will operate. It extends from the lowest to the highest value being measured.

Sensitivity—the ability of an instrument to detect small changes in a variable. It depends on the resolution of the display—the smallest change that can be read from the display.

Accuracy—the ability of an instrument to indicate the true value of a variable.

Calibration—the adjustment of an instrument so that its indicated values are as close as possible to their actual values.

Stability—the ability of an instrument to maintain accuracy over a certain time interval. A stable instrument seldom requires recalibration.

Frequency response—the range over which an instrument is able to follow changes in a variable. It is usually given in hertz (Hz) or cycles per second (cps).

Noise—unwanted variations in voltage that are mixed in with the signal of interest. It can be reduced to an acceptable level or eliminated by using filters.

1.4. Statistical Terminology.

In physiology, the results of repeated experiments will vary slightly, even if the experiments are performed under identical conditions. An isolated muscle, for example, when stimulated several times will contract to a slightly different length each time. Variations among different muscles will be even greater and must be recognized as a fact of life in making physiological measurements. This variation is called *natural variation*. Natural variation can be dealt with by elementary statistics. The following terms are commonly used to deal with natural variation:

Mean—the arithmetic average of a group of n numbers. To compute the mean, $\overline{x}$, add the numbers and divide by the number of measurements in the group. The mean is the value about which the individual measurements are distributed.

Range—the arithmetic difference between the lowest and the highest number in the group.

Mode—the value that occurs most frequently in a group of measurements.

Mean deviation—found by subtracting the mean from each measurement to derive a group
of mean deviations. The mean deviation is then found by adding the deviations and
dividing the sum by the number of deviations. Positive and negative signs may be
disregarded when summing the deviations. The mean deviation describes the average
amount of fluctuation of the measurements about their own mean.

Standard deviation—found by deriving a group of deviations from the mean, as
described above, and then squaring each deviation. The sum of the squared
deviations is then divided by $n - 1$. This result is called the variance; the
standard deviation is the square root of this variance. The standard deviation
is useful because 66-2/3 percent of the measurements will fall between the limits
$\overline{x} \pm 1$ S.D., 95 percent of the measurements will fall between the limits $\overline{x} \pm 2$ S.D.,
and 99 percent of the measurements will fall between $\overline{x} \pm 3$ S.D.

1.5. *t* Test of Significance. If we repeat a group of measurements over and
over, even the mean can be expected to vary from sample to sample. This variation
in sample means is said to occur "by chance alone." Thus, in order to decide if
two or more treatment groups are really different from each other or just appear
to be different due to chance, a statistical "test of significance" is needed.
There are many such tests available; for our purposes, a paired *t* test will be
used as an example of a test of significance. This test is limited to two groups
of "paired" or similar data.

Before applying a test of significance to a real situation, the level of significance that is acceptable must be decided. In most physiological situations,
the 5 percent level of significance is used. This means that the treatment groups
would be significantly different from each other on the basis of chance alone less
than 5 percent of the time.

We assume at the outset that there is no difference between Groups I and II.
This is called the *null hypothesis*.

GROUP I	GROUP II
2	3
1	4
2	5
3	4

The first step is to obtain a group of differences by subtracting the two groups
from each other:

GROUP I	GROUP II	DIFFERENCES
2	3	1
1	4	3
2	5	3
3	4	1

Once the differences are obtained, the original data are ignored and the differences
are used to complete the analysis. In the next step, the differences are summed to
give 8. This sum is referred to as D, and the average of the differences is $\overline{D} = 2$.

The next step is to calculate d as follows:

$$d = \text{(sum of each difference squared)} - \frac{D^2}{n}$$

where n is the number of differences.

$$d = (1^2 + 3^2 + 3^2 + 1^2) - \frac{8^2}{4} = 20 - 16 = 4$$

5

The variance (s^2) is found in the next step:

$$s^2 = \frac{d}{n-1} = \frac{4}{3} = 1.33$$

The standard deviation (s) is then found by taking the square root of the variance:

$$s = \sqrt{1.33} = 1.15$$

The standard deviation of the differences ($s_{diff.}$) is then given by:

$$s_{diff.} = \frac{1.15}{\sqrt{n}} = \frac{1.15}{2} = 0.577$$

Finally, t is calculated:

$$t = \frac{\overline{D}}{s_{diff.}} = \frac{2}{0.577} = 3.47$$

The t table (Table 1.2) is now consulted for $n - 1$ degrees of freedom, which is $4 - 1 = 3$ in our problem. Since the value of t calculated above (3.47) is larger than the value given in the table (3.18), there *is* a significant difference between these two groups at the 5 percent level of significance. As a result, the null hypothesis is rejected.

Experimental Procedure

In order to study physiological events, it is necessary to use a recording device that is sturdy, accurate, and reasonably simple to operate. Proceed to the section that applies to the equipment that you have available.

1.6. The Kymograph. This instrument consists of a cylindrical drum mounted on a shaft. The shaft rotates by means of a motor whose speed can be varied (Figure 1.1). The speed control knob should be turned to the desired setting only when the kymograph drum is turning. It is preferable to use as slow a speed as possible to avoid wasting paper and time.
To attach paper to the drum:

1. Remove the shaft and drum from the base.
2. Wrap the paper with its glazed surface outward around the drum and overlap it in the direction opposite to the direction of rotation.
3. Use tape to fasten the paper together at the top and bottom.
4. If the paper is narrower than the drum, secure the paper so that its lower edge is even with the lower edge of the drum.
5. Replace the shaft and drum on the base.

To adjust drum position:

1. Loosen the drum spring clamp.
2. Position the drum to the desired height.
3. Release the spring clamp.
4. Start recording at the top of the paper where it overlaps. After one revolution, adjust the height of the drum upward and continue recording.

Table 1.2 *t* Distribution at the 5 Percent and 1 Percent Levels of Significance

DEGREES OF FREEDOM	5%	1%
1	12.706	63.657
2	4.303	9.925
3	3.182	5.841
4	2.776	4.604
5	2.571	4.032
6	2.447	3.707
7	2.365	3.499
8	2.306	3.355
9	2.262	3.250
10	2.228	3.169
11	2.201	3.106
12	2.179	3.055
13	2.160	3.012
14	2.145	2.977
15	2.131	2.947
16	2.120	2.921
17	2.110	2.898
18	2.101	2.878
19	2.093	2.861
20	2.086	2.845
21	2.080	2.831
22	2.074	2.819
23	2.069	2.807
24	2.064	2.797
25	2.060	2.787
26	2.056	2.779
27	2.052	2.771
28	2.048	2.763
29	2.045	2.756
30	2.042	2.750
35	2.030	2.724
40	2.021	2.704
45	2.014	2.690
50	2.008	2.678
60	2.000	2.660
70	1.994	2.648
80	1.990	2.638
90	1.987	2.632
100	1.984	2.626
∞	1.960	2.576

Source: Table 1.2 is taken from Table III of Fisher & Yates: *Statistical Tables for Biological, Agricultural and Medical Research*, published by Longman Group Ltd. London (previously published by Oliver & Boyd Ltd. Edinburgh) and is used by permission of the authors and publishers.

7

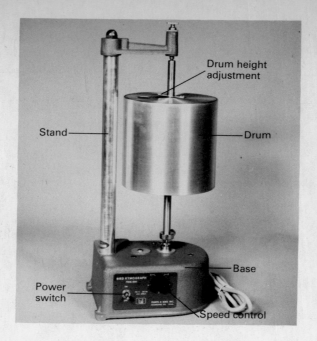

Figure 1.1   The kymograph.   (Photograph courtesy of Phipps and Bird, Inc.)

To record a physiological event:

1. Slide the kymograph up to the writing point(s).
2. The drum should barely touch the writing points to produce as little friction as possible.
3. Note the drum speed.
4. Measure the distance of the recorded event in convenient units.
5. Divide the distance of the recorded event by the drum speed to get the time course of the event.
6. Convert the distance of the recorded event to the appropriate units to get the amplitude of the event.

   1.7.   The Physiograph. This instrument consists of a set of ink-writing pens, a variable-speed paper drive, and provisions for various types of transducers and signal conditioners (Figure 1.2).
   To set up the physiograph:

1. Remove the plastic dust cover.
2. Plug in the main power cord.
3. Turn the tension wheel assembly so as to raise the pens, and remove the rubber cushion pad.
4. Place a paper towel under the pens. Raise the level of the inkwells to be used and gently squeeze the rubber bulb on each well while covering the hole in the bulb with your fingertip. The rate of ink flow can be controlled by raising or lowering the height of the inkwells—fast paper speeds require fast ink flows and vice versa.
5. Remove the paper towel and lower the pens onto the paper.
6. Open the front panel of the physiograph and see if there is enough paper. Make sure that each inkwell has enough ink.
7. Set the paper speed selector switch to the desired speed; set the time marker switch to the desired time interval.
8. Connect the ground wire to the ground terminal of the physiograph and to a cold water faucet.
9. Turn the main power switch on.

8

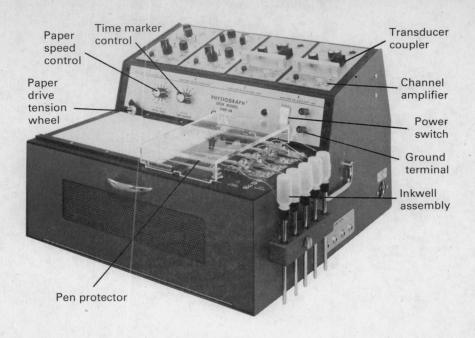

Figure 1.2 The desk–model physiograph. (Photograph courtesy of Narco Bio–Systems, Inc.)

10. Turn on each channel amplifier to be used as follows:

    a. Press the ready-record (red) button to the ready position.
    b. Turn on the channel amplifier button (white).
    c. Set the pens to appropriate baselines by turning the position knob(s).

11. Attach the transducers necessary for the experiment.

To shut down the physiograph:

1. Disengage the paper drive by moving the tension wheel lever.
2. Push the inkwells to their lowest level. Withdraw ink from the pens by squeezing the rubber bulb and then placing your finger over the hole in the bulb. When you release the bulb, keep your finger over the hole. Repeat this until you can see the ink withdraw.
3. Press all channel amplifier buttons off and turn off the main power switch.
4. Disconnect all transducers and transducer cables. Be careful not to bend or twist the connector pins; the ends of the transducer cables should be connected together to protect the pins.
5. Unplug the power cord and the ground wire.
6. Remove all ink stains from the physiograph with a moist paper towel.
7. Replace the rubber cushion pad and the plastic dust cover.

Figure 1.3   The stimulator.

   1.8.   The Stimulator.   This instrument is capable of delivering square-wave
pulses either as single stimuli or as trains of several stimuli (Figure 1.3).   For
our experiments, we will need to vary the amplitude (volts), duration (milliseconds),
and frequency (hertz) of stimuli.

To use the stimulator:

1.   Plug in the power cord and turn on the power switch.
2.   Set the output control switch for a monophasic or biphasic stimulus pulse.   For
     most of our experiments, a monophasic pulse will be used.
3.   The mode switch may be placed in one of three positions: off, single, or con-
     tinuous (CONT).   The single position will cause one stimulus pulse to be
     delivered; the CONT position will cause a train of several stimuli to be
     delivered.
4.   The monitor lamp flashes whenever a stimulus is delivered.
5.   The frequency controls select the stimulus repetition rate.   With the frequency
     multiplier switch, it is possible to vary the stimulus frequency from 0.2 to
     250 hertz.
6.   The duration controls select the stimulus width.   With the duration multiplier
     switch, it is possible to vary the stimulus duration from 0.1 to 120 milli-
     seconds.
7.   The voltage output controls select the stimulus amplitude.   With the output
     multiplier switch, it is possible to vary the stimulus output from 0 to 120
     volts.

   For today's experiment, connect the output jacks of the stimulator to the input
of the physiograph channel amplifier.   If you are using a kymograph instead, your
instructor will show you how to attach the output of the stimulator to the signal
magnet apparatus.   Try out each of the stimulator controls and observe its effect
on the recording.   Become familiar with each of the physiograph or kymograph controls.

After you have become familiar with the instrumentation, perform a *t* test on the following data:

GROUP I	GROUP II
16	12.0
23	23.0
26	23.0
23	21.0
23	19.5
22	15.5
17	13.0
19	17.5
19	17.0
21	15.0
30	19.5

Name _____ Lab Section _____ Date _____

t <u>Test</u>

Differences:

Sum of differences = D =

Average of differences = $\overline{D}$ =

$d = (\Sigma D^2) - \dfrac{D^2}{n}$ =

Number of differences = n =

Degrees of freedom = $n - 1$ =

$s^2 = \dfrac{d}{n - 1}$ =

$s = \sqrt{s^2}$ =

$s_{\text{diff.}} = \dfrac{s}{\sqrt{n}}$ =

$t = \dfrac{\overline{D}}{s_{\text{diff.}}}$ =

Conclusion: _____

_____

_____

EXPERIMENT 2

MEMBRANE TRANSPORT

Behavioral Objectives

The student should be able to:

1. State the relationship among osmotic pressure, concentration, the ideal gas constant, and absolute temperature.
2. Define the terms isotonic, isosmotic, hypotonic, and hypertonic.
3. Discuss the importance of the van't Hoff factor in calculation of the osmotic pressure.
4. State the relationship among freezing point depression, the van't Hoff factor, and concentration.
5. Perform standard tests for the observation of diffusion and osmosis.
6. Solve problems using formulae for osmotic pressure and freezing point depression.

Materials

Pan balances
Hotplates, with magnetic stirring feature
Magnetic stirring bars
Beakers (250 ml)
Test tubes (13 X 100)
Metal test tube holders
Blood sample
Test tube racks
Graduated cylinders (10 ml)
Hydrometers with reservoirs
Dialysis tubing (1-inch diameter)
Bromthymol blue indicator
Benedict's solution
Glucose (saturated)
NaCl (saturated)
Albumin (saturated)
Starch (boiled, 1 percent)
Iodine potassium iodide solution
Glycerol (1 M)
Ethylene glycol (1 M)
Urea (1 M)
Bunsen burner

15

Introduction

From the gas laws, it is known that $PV = nRT$, where P = pressure (atmospheres), V = volume (liters), n = number of moles of solute molecules, R = the ideal gas constant (0.082 liter-atmosphere/degree-mole), and T = absolute temperature (degrees Kelvin). If we rearrange this equation, we have:

$$P = \frac{nRT}{V} = \frac{n}{V}(RT)$$

The n/V term has the units of concentration (moles/liter) and hence concentration can be substituted for n/V in the equation to give $P = CRT$. This equation describes the behavior of solutions as well as gases. In general, this expression describes the *osmotic pressure* of a solution. Osmotic pressure is given the designation π to distinguish it from other types of pressure: $\pi = CRT$.

There are several terms used to compare one solution with another with respect to osmotic pressure. Keep in mind that there must always be *two* or more solutions that must be compared.

Isotonic—a solution that does not change in volume when separated from a test
   solution by a semipermeable membrane.
Isosmotic—a solution that has an osmotic pressure identical with that of the test
   solution. (Isotonic solutions are always isosmotic, but isosmotic solutions are
   not necessarily isotonic.)
Hypotonic—a solution that is more dilute (has more water) than the test solution.
Hypertonic—a solution that is more concentrated (has less water) than the test
   solution.

Recall that some electrolytes are completely ionized in water solution. NaCl, for example, behaves like two solute particles in solution. As a result, electrostatic forces among ions result in interactions that alter the actual number of ions present in a solution. For example, 1 mole of NaCl will behave like 1.8 or 1.9 moles of particles rather than 2 moles of particles. The degree of this interaction is a function of concentration—an infinitely small concentration of NaCl will approach the ideal case. The number that represents the actual number of particles in a solution is i, *the van't Hoff factor.*

Taking the van't Hoff factor into account in the equation derived above gives: $\pi = iCRT$. The value of i is different for each electrolyte; it is usually safe to assume a value for i equal to the number of particles produced by ionization of the electrolyte in question.

Since it is technically difficult to measure the osmotic pressure of a solution directly, an indirect method involving the freezing point depression is used. It is known that a 1 molar solution of a nonelectrolyte lowers the freezing point of an aqueous solution by 1.86°C. For example, a 1 molar sucrose solution freezes at -1.86°C. However, a 1 molar NaCl solution freezes at -3.48°C rather than at -3.72°C due to the van't Hoff factor. In general, the freezing point depression is given by $\Delta f_p = 1.86 iC$.

Experimental Procedure

2.1. Osmosis and Diffusion. Prepare four dialysis bags by cutting off four 6-inch sections of dialysis tubing. Immerse the bags in a beaker of distilled water until they are pliable enough to open. Then tie off one end of each bag with a strong knot. Fill the bags with the following solutions:

Bag 1—glucose (saturated)
Bag 2—starch (1 percent, boiled)
Bag 3—albumin (saturated)
Bag 4—NaCl (saturated)

Fill the bags until about 2 inches of solution is in each bag. Then tie off the other end of each bag, resulting in four "sausages" of solution. Blot the four bags on a paper towel and weigh each bag to the nearest 0.1 gram on a pan balance. Enter these initial weights in the table on the data sheet.
Place each bag in a separate beaker and add 150 ml of distilled water to each beaker. Allow the bags to equilibrate in the beakers for one hour. Place the beaker containing Bag 4 (NaCl) on a magnetic stirrer and stir the solution during its equilibration. When the equilibration has been completed, remove each bag, blot it with a paper towel, and reweigh it. Enter these final weights in the table on the data sheet. Then subtract the initial weights from the final weights to give the weight change for each bag.

1. Which bags gained water by osmosis?

To determine if diffusion has occurred, test the water remaining in the beakers as follows:

Beaker 1: Place 10 drops of water from beaker 1 in a test tube. Add 10 drops of Benedict's solution, and place the test tube in a boiling water bath for 5 minutes. If the contents of the tube turn a yellow-green color, it indicates a faint trace of glucose; a bright orange-red color indicates a sizable amount of glucose.

2. Is glucose present in the water?

Beaker 2: Place a drop of water from beaker 2 on a microscope slide. Add a drop of IKI solution. If starch is present, a dark blue-black color will result; if no starch is present, a faint yellow color will result.

3. Is starch present in the water?

Beaker 3: Fill a test tube about two-thirds full of water from beaker 3. Hold the tube with a test tube holder and place the upper third of the tube in a Bunsen burner flame. If a white cloud appears, albumin is present; if no cloud appears, albumin is not present.

4. Is albumin present in the water?

Beaker 4: Place a sample of water from beaker 4 in a hydrometer reservoir. Then place the hydrometer float in the reservoir. It is important that the hydrometer freely floats in the water. Read the specific gravity of the water at the water-air interface. A specific gravity greater than 1.000 indicates the presence of solute (NaCl in this case).

5. Is NaCl present in the water?

6. Why were some substances unable to diffuse out of the bags?

2.2. Diffusion. Fill another dialysis bag with tap water and add enough bromthymol blue indicator to turn the water a light blue color. If your bag does not turn blue, empty it and fill it with tap water from a different faucet. Tie off the top of the bag and immerse it in a beaker containing 150 ml of 0.1-N HCl. Allow the bag to equilibrate for 2 minutes.

7. What is the appearance of the bag after 2 minutes?

8. Why does the bromthymol blue indicator change color?

2.3. Hemolysis Rates. Obtain three test tubes and place them in a test tube rack. Put 2 ml of urea in the first tube, 2 ml of ethylene glycol in the second tube, and 2 ml of glycerol in the third tube. Note the time, and then add 2 drops of blood to the first tube. Determine how long it takes for the solution to become transparent red. Repeat the procedure for the second and third tubes. Enter your times on the data sheet.

9. Why do the urea and ethylene glycol tubes become transparent rapidly?

10. Why does the glycerol tube remain cloudy for a long time?

2.4. Problems. Solve the following problems based on osmotic pressure and freezing point depression:

1. What is the osmotic pressure of a solution of 300 g of sucrose in 1 liter of water at 25°C? What is the osmotic pressure at 30°C? The molecular weight of sucrose is 342 and assume that for sucrose $i = 1$.

2. What is the freezing point depression of a solution with an osmotic pressure of 4.58 atm at 0°C? The van't Hoff factor for the solute in this solution is 1.

3. What is the osmotic pressure of an 0.4 molar solution of sugar at 25°C? Assume that for the sugar $i = 1$.

4. A sugar solution freezes at the same temperature as an 0.3 molar NaCl solution. If the van't Hoff factor for NaCl is 1.82, what is the concentration of the sugar solution?

5. An 0.4 molar solution of KCN freezes at −0.75°C. Find i and the osmotic pressure at 5°C. Repeat the calculations for an 0.4 molar solution of H_2SO_4 which freezes at −1.75°C.

6. The average osmotic pressure of human blood is 7.7 atm at 40°C. What is the total concentration of the solutes found in the blood? If you assume that this concentration is equal to the molarity, find the freezing point of blood.

7. A cell is isotonic to an 0.45 molar solution of sucrose and also with an 0.36 molar solution of a monovalent salt. What is the van't Hoff factor of the monovalent salt?

8. A sample of urine freezes at −0.69°C. What is its osmotic pressure at 25°C? With what molar concentration of sucrose would this be isotonic? If you assume $i_{NaCl} = 1.82$, with what molar concentration of NaCl would this be isotonic?

DATA SHEET FOR EXPERIMENT 2:   MEMBRANE TRANSPORT

Name_____ Lab Section_____ Date _____

2.1. Osmosis and Diffusion

Bag	Initial Wt. (g)	Final Wt. (g)	Weight Change (g)
1 (Glucose)			
2 (Starch)			
3 (Albumin)			
4 (NaCl)			

1. Which bags gained water by osmosis? _____

2. Is glucose present in the water? _____

3. Is starch present in the water? _____

4. Is albumin present in the water? _____

5. Is NaCl present in the water? _____

6. Why were some substances unable to diffuse out of the bags?

    _____

    _____

    _____

2.2. Diffusion

7. What is the appearance of the bag after 2 minutes?

    _____

8. Why does the bromthymol blue indicator change color?

    _____

    _____

    _____

21

2.3. Hemolysis Rates

Tube	Time to Transparency (sec)
1 (Urea)	
2 (Ethylene glycol)	
3 (Glycerol)	

9. Why do the urea and ethylene glycol tubes become transparent rapidly?

_____

_____

10. Why does the glycerol tube remain cloudy for a long time?

_____

_____

_____

FROG NEUROPHYSIOLOGY

Behavioral Objectives

The student should be able to:

1. Explain how a biphasic action potential is recorded from a nerve.
2. Explain why a compound action potential appears graded rather than all-or-none.
3. Construct a strength-duration curve.
4. Demonstrate the phenomena of threshold and latency.
5. Calculate the conduction velocity of a nerve.
6. Measure the refractory period of a nerve.
7. Demonstrate the effects of low calcium and anesthetics on nerve excitability.

Materials

Bullfrogs
Wax dissecting pans
Dissecting pins
Dissecting instruments
Fine thread
Normal frog Ringer
Low calcium frog Ringer
Novocaine (2 percent)
Rulers (mm)
Storage oscilloscope, with differential amplifier
Stimulator, with stimulus isolation unit
Sciatic nerve chamber

Introduction

All nerves are made up of nerve cells called *neurons* that have three parts: (1) a *soma* (cell body), which contains numerous cell organelles; (2) a system of *dendrites*, short cytoplasmic processes that conduct electrical activity toward the soma; and (3) an *axon*, a long cytoplasmic process that conducts electrical activity away from the soma.

The *action potential* is the basic unit of electrical activity in the nervous system. The most common method for studying the action potential involves removing a neuron or neurons from an experimental animal, passing electrical current across the neuron cell membrane, and recording the resulting changes in membrane potential.

The changes in membrane potential recorded with external electrodes are complicated in appearance. As shown in Figure 3.1, the membrane potential is recorded with one polarity (upward) when the action potential passes the first recording electrode (1), the membrane potential is zero when the action potential is between the two recording electrodes (2), and the membrane potential is recorded with the opposite polarity (downward) when the action potential passes the second recording electrode (3). This type of recording is called a *diphasic action potential*.

It is possible to produce a monophasic action potential by preventing the electrical activity from reaching the second recording electrode. This can be done by physical damage to the nerve membrane (crushing), removal of the resting potential (using high K^+ solutions), or by blocking depolarization (using local anesthetics).

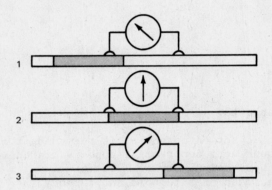

Figure 3.1 Recording of action potential by external electrodes.

Nerves are made up of several populations of axons rather than a single axon. As a result, action potentials recorded from whole nerves are further complicated by two factors. First, the action potential amplitude is graded with stimulus intensity. In each axon of the nerve, the action potential is all-or-none, but as more and more axons are depolarized to threshold by increases in stimulus intensity, the action potential amplitude increases until all of the axons have been stimulated. Second, the axons within a nerve exhibit differences in conduction velocity as a result of differences in axon diameter and myelination. As a result, the action potential may show humps in its falling phase corresponding to these different groups of axons. Hence action potentials recorded from whole nerves are called *compound action potentials*.

Experimental Procedure

In this experiment, we will examine the electrical properties of the bullfrog sciatic nerve. Kill a bullfrog by double pithing it, and cut its skin transversely around the chest just behind the forelegs (Figure 3.2 a, b, and c). Strip the frog of its skin by gripping the skin with a pair of heavy forceps and by pulling the skin down to its feet. Place the frog in a wax dissecting pan with its dorsal side facing upward. The sciatic nerve is buried in the middle of the thigh and is white in color. Carefully separate the thigh muscles with a dissecting needle until the nerve is identified. Free as much of the nerve as possible from the surrounding muscle. Insert the tips of a fine forceps under the nerve and use them to draw a piece of thread under the nerve. Slide the thread to one end of the nerve and tie it with a strong knot. After unburying several inches of the nerve, pass another thread under the nerve and tie it on the other end as well. Then remove the nerve by cutting it beyond the knots. Lay the sciatic nerve horizontally across the

(a)

Figure 3.2 Dissection of frog sciatic nerve. (a) Frog with dorsal surface facing
upward.

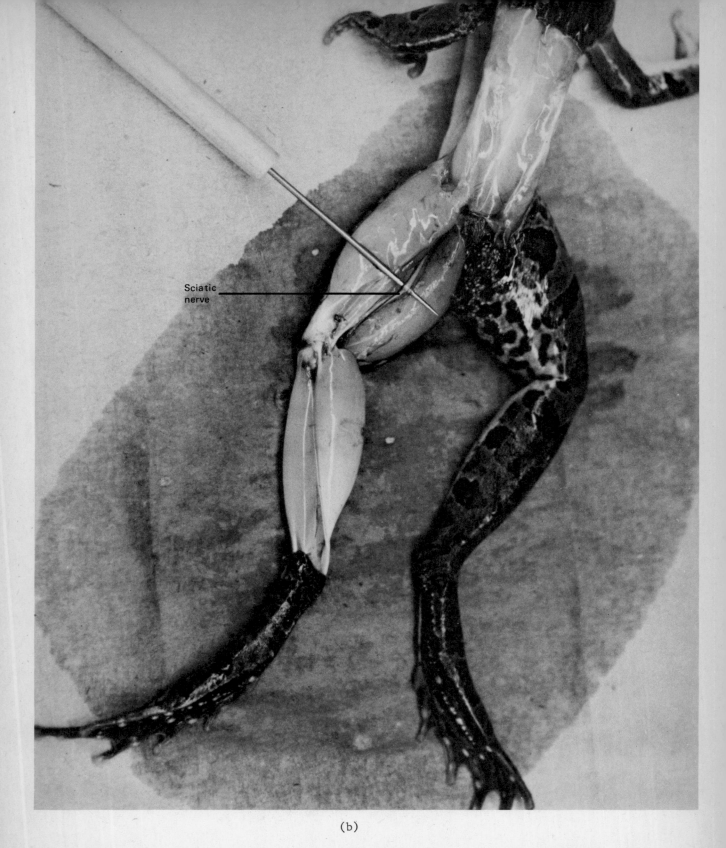

Sciatic
nerve

(b)

Figure 3.2   (Continued) Dissection of frog sciatic nerve. (b) Location of sciatic
             nerve.

(c)

Figure 3.2 (Continued) Dissection of frog sciatic nerve. (c) Sciatic nerve tied on
both ends.

silver-silver chloride wires of the nerve chamber. Be careful not to stretch the nerve when mounting it. Connect the external stimulating electrodes to the output of the stimulus isolation unit, and connect the external recording electrodes to the input of the storage oscilloscope as shown in Figure 3.3 a and b.

During the dissection and throughout the experiment, keep the nerve moist with frog Ringer solution. This solution will prevent the nerve from dying. Since the blood supply to the nerve has been interrupted by the dissection procedure, the constant exchange between the tissue fluid around the cells and the plasma has stopped. To substitute for some of the normal functions of the circulatory system, Ringer must be added to the nerve to prevent the osmotic pressure of the intracellular fluids from increasing due to dessication. The Ringer solution, developed by Sidney Ringer in the 1800s, contains a buffered ionic solution that is isotonic to frog tissue fluid. It is important to note that physiological saline solutions have been developed for other experimental animals but are not to be confused with Ringer solution which refers specifically to frog.

Because of the design of the nerve chamber, it is important to keep the level of the Ringer solution below that of the silver wires. If the Ringer comes in contact with the wires, the stimulus pulse will be shorted out by traveling through the Ringer rather than through the nerve.

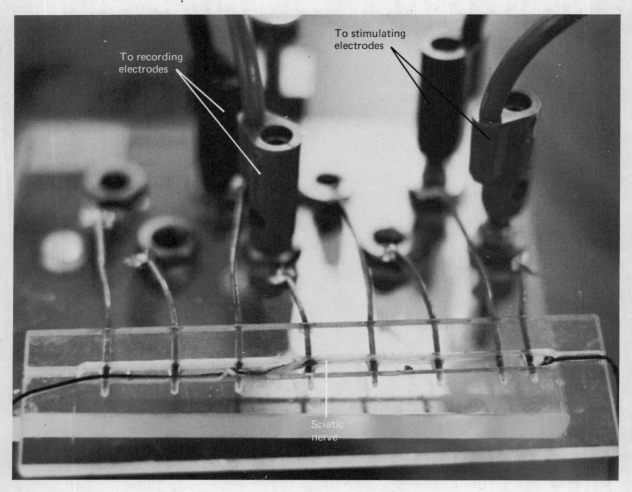

(a)

Figure 3.3 Frog sciatic nerve apparatus. (a) Nerve mounted in chamber.

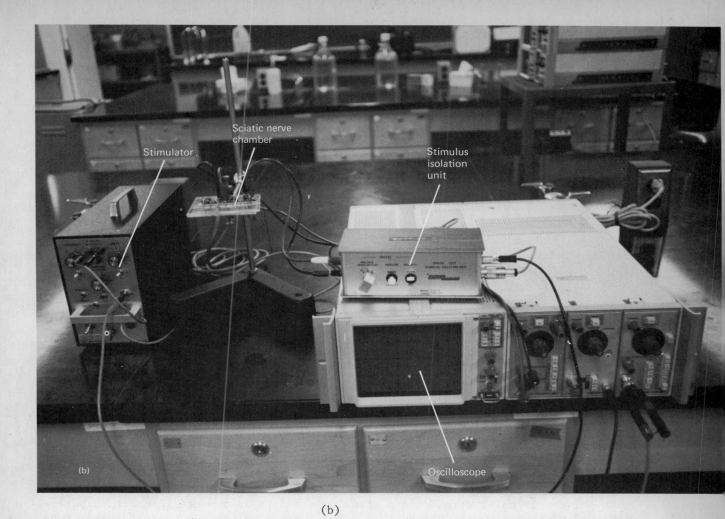

(b)

Figure 3.3 (Continued) Frog sciatic nerve apparatus. (b) Instrumentation for
stimulation, conditioning, and display.

3.1. Action Potential Threshold. A stimulus that does not cause an action
potential is referred to as *subthreshold*. The minimum stimulus that causes an
action potential is said to be *threshold*. When the stimulus intensity is large
enough so that all of the axons in the nerve are firing action potentials, the
stimulus is termed *maximal*. Any stimulus with an intensity greater than maximal
is called *supramaximal*.

To determine the action potential threshold, turn the stimulus voltage to a
low value and set its duration for 25 msec. Deliver a single, biphasic stimulus
to the nerve and observe the tracing on the oscilloscope screen. Gradually increase the stimulus voltage and repeat until a small action potential is visible
on the screen.

1. What is the action potential threshold?

Continue increasing the stimulus voltage until no further increase in action
potential amplitude is noted.

29

2. What is the maximal stimulus voltage?

The oscilloscope amplifier setting and the amplitude of the action potential tracing can be used to calculate the actual amplitude of the action potential in millivolts.

3. What is the maximal action potential amplitude?

Rinse the nerve with Ringer solution and allow it to rest for a minute. Then repeat the above procedure.

4. What is the mean threshold voltage?

5. What is the mean maximal stimulus voltage?

6. What is the mean maximal action potential amplitude?

By changing the duration of the stimulus, it is possible to study the relationship between stimulus duration and stimulus strength in producing an action potential. Find the stimulus voltage that gives the maximal action potential amplitude for stimulus durations of 50, 100, and 120 msec. Construct a strength-duration curve by plotting stimulus strength (volts) versus stimulus duration (milliseconds). Your graph should be similar to Figure 3.4.

7. What general statement can be made from this graph?

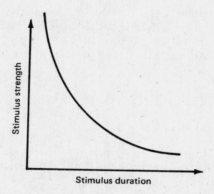

Figure 3.4   The strength-duration curve.

3.2. Compound Action Potential. Change the settings on your oscilloscope so that the action potential will be displayed with a large amplitude (increase the amplifier sensitivity) and a wider time course (decrease the time base setting). Deliver a single supramaximal stimulus to the nerve and study the different parts of the tracing obtained. Try to identify the following parts of the action potential:

Stimulus artifact—produced by the stimulating electrodes during each stimulus.
Latency—the time from the end of stimulation to the beginning of the action potential.
Foot—the time during which the axons are depolarizing to threshold.
Rising phase—the time when Na$^+$ ions flow into the axons and depolarize the membranes.
Overshoot (peak)—the time when the insides of the axons become positively charged.
Falling phase—the time when K$^+$ ions flow out of the axons and repolarize the membranes.
Afterpotentials—prolonged depolarizations or hyperpolarizations at the end of the falling phase.

8. Have you noticed any humps in the falling phase of the action potential? If so, do they disappear with an increase in stimulus intensity?

9. In which direction (left-to-right or right-to-left) is the action potential moving in your nerve?

10. If you stimulated the nerve in the center, in which direction(s) would the action potential move? Why?

3.3. Conduction Velocity. Measure the distance from the first stimulating electrode to the first recording electrode with a millimeter ruler. Then deliver a supramaximal stimulus to the nerve. Measure the time in milliseconds from the beginning of the stimulus artifact to the foot of the action potential. The conduction velocity can be calculated by the following equation:

C.V. = distance/time = mm/msec = m/sec

11. What is the conduction velocity of your nerve?

3.4. Refractory Period. If you have a stimulator that is capable of producing two stimuli with a variable delay between them, it is possible to determine the refractory period of the nerve. Set the intensity of both stimuli to a supramaximal voltage. By applying two stimuli to the nerve, two action potentials will be initiated and recorded in the nerve. Increase the duration of the first stimulus and note its effect on the second action potential. The second action potential should become smaller and gradually disappear as the second stimulus begins during the refractory period of the first action potential.

31

12. What happens to the amplitude of the second action potential in your nerve?

13. What is the refractory period of your nerve in milliseconds?

    3.5. Effect of Low Calcium. Rinse the nerve with several drops of low calcium Ringer solution. Wait for 1 minute and then determine the new action potential threshold.

14. What is the new threshold voltage?

    Rinse the nerve with normal Ringer solution and wait for 3 minutes before proceeding.

    3.6. Effect of Local Anesthetics. Apply a few drops of novocaine in the vicinity of the recording electrode farthest from the stimulating electrodes. Wait 1 minute and then apply a supramaximal stimulus to the nerve. Compare your recording with those shown in Figure 3.5.

15. Is your recording biphasic or monophasic?

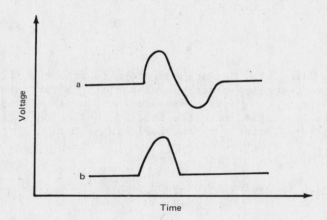

Figure 3.5 Sciatic nerve compound action potential. (a) Biphasic. (b) Monophasic.

32

DATA SHEET FOR EXPERIMENT 3: FROG NEUROPHYSIOLOGY

Name _____ Lab Section _____ Date _____

3.1. Action Potential Threshold

1. What is the action potential threshold? _____

2. What is the maximal stimulus voltage? _____

3. What is the maximal action potential amplitude? _____

4. What is the mean threshold voltage? _____

5. What is the mean maximal stimulus voltage? _____

6. What is the mean maximal action potential amplitude? _____

Duration (msec)	Stimulus voltage (V)
25	
50	
100	
120	

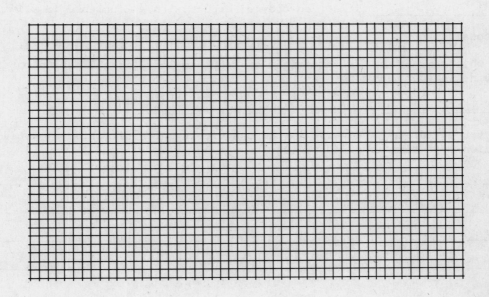

7. What general statement can be made from this graph?

_____

_____

_____

3.2. Compound Action Potential

Sketch of biphasic compound action potential:

$$mV \llcorner$$
msec

8. Have you noticed any humps in the falling phase of the action potential?

_____ If so, do they disappear with an increase in stimulus

intensity? _____

9. In which direction (left-to-right or right-to-left) is the action potential
moving in your nerve?

_____

10. If you stimulated the nerve in the center, in which direction(s) would the
action potential move? Why?

_____

_____

_____

_____

3.3. Conduction Velocity

11. What is the conduction velocity of your nerve?

_____

3.4. Refractory Period

12. What happens to the amplitude of the second action potential in your nerve?

_____

13. What is the refractory period of your nerve in msec?

_____

3.5. Effect of Low Calcium

14. What is the new threshold voltage? _____

3.6. Effect of Local Anesthetics

15. Is your recording biphasic or monophasic?

_____

EXPERIMENT 4

SENSORY RECEPTORS AND HUMAN REFLEXES

Behavioral Objectives

The student should be able to:

1. State the classification scheme used for sensory receptors.
2. Demonstrate the difference in distribution of touch, pressure, and temperature receptors.
3. Demonstrate the phenomena of adaptation and referred pain.
4. Elicit normal human reflexes used as signs of normal body function.

Materials

Knee reflex hammers
Plastic rulers (mm)
Fine thread
Microscope slides
Lens paper
Meter sticks
Hotplates
Pins
Ice buckets
Beakers (1 liter)
Beakers (250 ml)
Alcohol swabs
Calipers

Introduction

Information about the internal and external environment reaches the central nervous system via a variety of *sensory receptors*. These receptors act as transducers which convert various stimuli into nerve action potentials. Receptors exist in the body for the conversion of mechanical, thermal, electromagnetic, and chemical stimuli.

Traditionally, receptors have been classified as cutaneous or special. The cutaneous receptors respond to touch, pressure, heat, cold, and pain. The special sense receptors respond to smell, taste, vision, and hearing. In this experiment, we will confine ourselves to the cutaneous sensations; in the next experiment, we will consider the special sense receptors.

Other classification schemes have been used to classify receptors into: (1) *teleceptors*, those receptors concerned with events at a distance; (2) *exteroceptors*, those receptors concerned with the external environment of the body; (3) *interoceptors*, those receptors concerned with the internal environment of the body; and (4) *proprioceptors*, those receptors that provide information about body position. However, because pain is initiated by noxious stimuli, pain receptors are sometimes classified as *nociceptors*. In addition, the term *chemoreceptor* is used to refer to receptors stimulated by a change in their chemical environment.

The basic unit of response to sensation is the reflex arc. A *reflex arc* consists of a sensory receptor, an afferent neuron, one or more synapses in a central integrating center, an efferent neuron, and an effector. The simplest reflex arc is one with a single synapse between the afferent and efferent neurons in the integrating center. Such reflex arcs are termed *monosynaptic*. Reflex arcs in which one or more interneurons are interposed between the afferent and efferent neurons are termed *polysynaptic*. The number of synapses in a polysynaptic reflex arc varies from two to several hundred, depending on the reflex.

Experimental Procedure

4.1. **Touch Receptors.** Receptors for touch vary in their distribution and density over different body surfaces. Areas of the body having numerous touch receptors have a finer sense of "feel" than areas that have only a few touch receptors.

Have your partner sit on the lab bench with eyes closed. Use a caliper to apply tactile stimuli to your partner's skin. Start with the points of the caliper far apart and gradually decrease the distance between the points until your partner feels only one point. Be careful to apply both points of the caliper simultaneously each time. Record the two-point threshold by measuring the space between the caliper points with a millimeter ruler. Record your results in the table on the data sheet for the following body areas:

1. Fingertip
2. Palm of hand
3. Back of forearm
4. Back of neck
5. Calf of leg
6. Tongue (Clean caliper with an alcohol swab first!)

Then have your partner close his eyes and touch him with one point of the caliper in the areas listed above. After removing the point of the caliper, have your partner try to touch the same spot with the point of a pen. Measure the distance from the point of stimulation with the millimeter ruler and enter the error distance in the table on the data sheet. Then choose one of the areas where localization was poor (large error distance). Repeat the localization procedure three times using the same stimulation point.

<u>1.</u> <u>Is localization improved with repeated trials?</u>

4.2 **Pressure Receptors.** Obtain a glass microscope slide, a booklet of lens paper, and a hair from your head. Place the hair on top of the glass slide on the lab bench, and cover the hair with several sheets of lens paper. (Do not tear the lens paper out of the booklet!) Move the tip of your index finger across the paper so as to feel the hair.

2. What is the maximum number of pages through which you can feel the hair?

    Repeat the procedure using your knuckle and the edge of your palm.

3. Which of the three body surfaces contains the most pressure receptors?

    Using a pencil, move one hair on your partner's forearm. Move the hair as slowly as possible until it springs away from the pencil. The slow bending should lead to rapid adaptation of the receptors and should not be felt. The rapid change when the hair springs back should be felt.

4. Is the sensation greater when the hair is slowly bent or when it springs away?

5. What is the function of receptor adaptation?

    4.3. Temperature Receptors. Prepare three 250-ml beakers with the following contents:

1. Ice water (0 to 5°C)
2. Room-temperature water (20 to 25°C)
3. Hot water (40 to 45°C)

Place your left index finger in the ice water and your right index finger in the hot water for 2 minutes.

6. What happens to the sensations of cold and warmth in each finger after 2 minutes?

7. Which finger adapts fastest?

    Then place both fingers in the beaker containing room-temperature water.

8. What are the sensations in each finger?

4.4. __Pain Receptors__. Referred pain is a phenomenon in which pain is perceived in one body area when another body area is actually receiving the painful stimulus. The pain is "referred" to the other more remote area. Have your partner place his elbow in a beaker of ice water for a few minutes.

__9.__ __Is any change noted in the location of the pain sensations?__

__10.__ __If so, where is the referred pain felt?__

The mediator for this referred pain sensation is the ulnar nerve which passes over the elbow joint on its way to innervating the palm and fingers.

4.5. __Sneezing Reflex__. Tickle the interior of your partner's nostril with a piece of fine thread.

__11.__ __What is the effect?__

__12.__ __What is the function of this reflex?__

4.6. __Pupil Reflex__. Have your partner face the window so that both eyes are equally illuminated. Then have your partner close the eyes and cover them with both hands for 30 seconds. Then, expose the eyes to room light.

__13.__ __Is there a light reaction in the pupils?__

4.7. __Consensual Pupil Reflex__. Repeat the above experiment but cover only one eye for 30 seconds. Watch the uncovered eye when the covered eye is exposed to light.

__14.__ __Does the size of both pupils change?__

4.8. __Accommodation Reflex__. Have your partner hold a pin at arm's length and have him or her focus on it. Slowly bring the pin toward the eyes while keeping the tip in focus until it appears double. Then move the pin back to a point where it appears as one pin again. This is the near point for accommodation. Measure the near point in centimeters by using a meter stick. Compare your value with the following normal values:

AGE	NEAR POINT (cm)
10	7
20	10
30	14
40	22
50	40
60	100

15. What is accommodation?

16. What eye muscles are involved in accommodation?

4.9. Oculo-Cardiac Reflex. Have your partner close his eyes and count his pulse rate for 15 seconds. Multiply this count by four to obtain the pulse rate per minute. Then have your partner gently apply pressure on both eyeballs with his fingertips. Count the pulse rate for 15 seconds and multiply by four to obtain the new pulse rate.

17. What is the resting pulse rate?

18. What is the pulse rate after pressure is applied?

19. Did the pulse rate increase, decrease, or remain the same?

4.10. Cilio-Spinal Reflex. Seat your partner on the lab bench. Then pinch the skin on the back of your partner's neck while watching his eye pupils.

20. Do the pupils change size? If so, do they constrict or dilate?

4.11. Patellar (Knee Jerk) Reflex. Have your partner sit on the lab bench with her knees flexed and her legs completely relaxed. Strike the patellar tendon with a reflex hammer. This causes a stretching of the quadriceps femoris muscle.

41

21. What is the response?

    Absence of the knee jerk reflex may indicate disease of the nerve to or from
the quadriceps femoris muscle or in the integrating center for this reflex in the
spinal cord. Repeat the procedure with your partner gritting her teeth and
clenching her fists.

22. Is there any change in the magnitude of the reflex?

23. If so, is the response greater or lesser than before?

    4.12. Achilles Tendon (Ankle Jerk) Reflex. Have your partner kneel down and
flex his ankle to increase tension on the gastrocnemius muscle in the lower leg.
Strike the Achilles tendon with a reflex hammer.

24. Describe the response seen and felt in the gastrocnemius muscle.

    Absence of the ankle jerk reflex may indicate disease in the lumbosacral region
of the spinal cord. This reflex is also used as the basis of a test for thyroid
gland function. The contraction phase is normal in all thyroid states, but the
relaxation phase is prolonged in hypothyroidism and accelerated in hyperthyroidism.

    4.13. Plantar Reflex. Remove the sock from your partner's foot. Run a blunt
instrument (the handle of the reflex hammer) along the sole of the foot from the
heel to the toes, as shown in Figure 4.1.

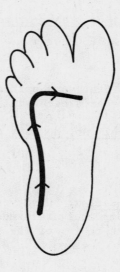

Figure 4.1

42

<u>25.</u> <u>What is the response?</u>

The normal adult response is plantar flexion of the toes. In people with pyramidal tract lesions, the response is dorsiflexion of the toes (a positive Babinski sign). However, it is normal for children under 18 months of age to show a positive Babinski sign because of incomplete maturation of the nervous system.

   <u>4.14.</u> <u>Withdrawal Reflex</u>. Have your partner sit at the lab bench and close his eyes, with the palm of one hand facing up on the lab bench. Then, using a pin cleaned with an alcohol swab, suddenly touch your partner's index finger.

<u>26.</u> <u>What happens to the forearm?</u>

DATA SHEET FOR EXPERIMENT 4: SENSORY RECEPTORS AND HUMAN REFLEXES

Name _____ Lab Section _____ Date _____

4.1. Touch Receptors

Area	Two-Point Threshold (mm)
Back of neck	
Fingertip	
Palm of hand	
Back of forearm	
Tongue	
Calf of leg	

Area	Error Distance (mm)
Back of neck	
Fingertip	
Palm of hand	
Back of forearm	
Calf of leg	

1. Is localization improved with repeated trials? _____

4.2. Pressure Receptors

2. What is the maximum number of pages through which you can feel the hair?

    Finger_____ Knuckle_____ Palm_____

3. Which of the three body surfaces contains the most pressure receptors?

    _____

4. Is the sensation greater when the hair is slowly bent or when it springs away?

    _____

5. What is the function of receptor adaptation?

_____

_____

_____

_____

4.3. Temperature Receptors

6. What happens to the sensations of cold and warmth in each finger after 2 minutes?

Right finger _____

Left finger _____

7. Which finger adapts fastest? _____

8. What are the sensations in each finger?

Right finger _____

Left finger _____

4.4. Pain Receptors

9. Is a change noted in the location of the pain sensations?

_____

10. If so, where is the referred pain felt? _____

_____

4.5. Sneezing Reflex

11. What is the effect? _____

12. What is the function of this reflex? _____

4.6. Pupil Reflex

13. Is there a light reaction in the pupils? _____

4.7. Consensual Pupil Reflex

14. Does the size of both pupils change? _____

4.8. Accommodation Reflex

Near point _____ cm

15. What is accommodation? _____

_____

16. What eye muscles are involved in accommodation?

_____

_____

4.9. Oculo-Cardiac Reflex

17. What is the resting pulse rate? _____

18. What is the pulse rate after pressure is applied?

_____

19. Did the pulse rate increase, decrease, or remain the same?

_____

4.10. Cilio-Spinal Reflex

20. Do the pupils change size? _____ If so, do they

constrict or dilate? _____

4.11. Patellar (Knee Jerk) Reflex

21. What is the response? _____

22. Is there any change in the magnitude of the reflex?

_____

23. If so, is the response greater or lesser than before?

_____

4.12. Achilles Tendon (Ankle Jerk) Reflex

24. Describe the response seen and felt in the gastrocnemius muscle._____

_____

_____

4.13. Plantar Reflex

25. What is the response? _____

_____

_____

4.14. Withdrawal Reflex

26. What happens to the forearm? _____

_____

EXPERIMENT 5

HUMAN SPECIAL SENSES

Behavioral Objectives

The student should be able to:

1. Map the location of taste buds on the tongue.
2. State which cranial nerves carry taste sensations to the brain.
3. Determine the approximate taste threshold for sucrose.
4. Demonstrate the interaction between taste and smell sensations.
5. Demonstrate olfactory adaptation.
6. State which cranial nerve carries smell sensations to the brain.
7. Demonstrate bone conduction in hearing.
8. Perform the Weber and Rinne tests for hearing loss.
9. Demonstrate the presence of the blind spot on the retina.
10. Calculate visual acuity by using a Snellen eye chart.
11. Demonstrate the phenomenon of an afterimage.
12. Perform color blindness tests using the Ishihara books.
13. Demonstrate the phenomenon of binocular vision.

Materials

Apple, potato, carrot, onion
Cotton-tipped wood applicator sticks
Phenylthiocarbamide (PTC) paper
Spatulas
Powdered sucrose
Sucrose solutions (1:1000, 1:500, 1:250, 1:100, 1:10)
Quinine sulfate (0.1 percent)
Acetic acid (1 percent)
NaCl (10 percent)
Methyl salicylate (wintergreen oil)
Tuning forks (512 Hz)
Snellen eye chart
Ishihara color blindness test books
Sterile cotton
Alcohol swabs
Index cards (3 X 5)
Test tubes (large size)
Tap water in squeeze bottles
White paper (four sheets)
Colored tape (red, green, yellow)

Introduction

The special senses include *gustation* (taste), *olfaction* (smell), *vision*, and *hearing*. Each of these senses is "specialized" to receive different stimuli: gustatory and olfactory cells respond to chemicals dissolved in the mouth or air, respectively; rods and cones respond to light; hair cells respond to pressure waves.

The receptor for the sense of taste is the taste bud. In an average adult, there are about 10,000 taste buds which are widely and unevenly distributed on the tongue, soft palate, and pharynx. Each taste bud is sensitive to all tastes but not to the same degree. It has been suggested that the brain interprets taste by a statistical comparison of which taste buds are responding most vigorously at a given time.

The receptors for smell are found in the olfactory epithelium of the nose. The sense of smell is the least understood of the four special senses. It is thought that mixtures of the primary smell sensations produce the broad spectrum of odors with which we are familiar.

The receptor for vision is the eye. Visible light is perceived as it is transmitted through or reflected from objects. The human visual range extends from the short violet wavelengths to the long red wavelengths of light. Ultraviolet and infrared wavelengths of light are invisible to man. The rods and cones, found in the retina of the eye, are sensitive to light. Rods are sensitive to low light intensities and are used for black-and-white vision. Cones are sensitive to high light intensities and are used for color vision and visual acuity.

The receptor for hearing is the ear. Sound waves must be conducted through a series of structures to the cochlea of the inner ear. Hair cells are present in the cochlea which respond to bending caused by fluid pressure waves.

Experimental Procedure

5.1. Gustation. Dry your partner's tongue with a tissue and, using a spatula, place a few grains of powdered sucrose on your partner's tongue.

1. Is the sucrose tasted immediately? Why?

To map the location of taste buds that respond most strongly to different tastes, the following solutions will be used:

Sucrose (1:10)—sweet
Acetic acid (1 percent)—sour
NaCl (10 percent)—salty
Quinine sulfate (0.1 percent)—bitter

Have your partner stick his tongue out and close his eyes. Dry the tongue with a tissue. Moisten a cotton-tipped applicator stick with one of the above solutions and apply the swab to the tip, sides, middle, back, and underside of the tongue. Use the quinine sulfate solution last to prevent a bitter aftertaste from interfering with the results. Record the taste perceived for each area in the space provided on the data sheet. After each solution has been studied, wash the tongue and dry it.

<u>2.</u> <u>Which cranial nerves carry taste sensations to the brain?</u>

To determine the threshold for taste sensation, have your partner close her eyes. Place a drop of 1:1000 sucrose solution on the tip of her tongue. If it is not perceived as sweet, rinse the mouth and repeat with sucrose solutions of 1:500, 1:250, 1:100, and 1:10 until the taste threshold has been reached. Taste thresholds vary for different people and for different substances.

<u>3.</u> <u>Which sucrose solution was closest to the taste threshold for your partner?</u>

To demonstrate the interaction between taste and smell, food cubes will be prepared from the following foods:

1. Apple
2. Potato
3. Carrot
4. Onion

Have your partner close her eyes and hold her nostrils closed. Dry her tongue and place one of the above food cubes on her tongue.

<u>4.</u> <u>Can your partner identify the food cube?</u>

Then allow your partner to chew the food cube.

<u>5.</u> <u>Can your partner identify the food cube after chewing?</u>

Finally, allow your partner to open her nostrils.

<u>6.</u> <u>Can your partner identify the food cube with open nostrils?</u>

The ability to taste certain substances is inherited. Phenylthiocarbamide (PTC) is an example of such a substance. PTC produces a bitter taste in about 70 percent of the population, whereas about 30 percent of the population experience no taste at all. Place a strip of paper impregnated with PTC on your tongue and then chew it.

<u>7.</u> <u>Can you taste PTC?</u>

5.2. <u>Olfaction</u>. To demonstrate olfactory adaptation, have your partner close his eyes and press one nostril closed. Hold a cotton applicator stick that has been dipped in wintergreen oil (methyl salicylate) under his open nostril.

<u>8. Is the smell detected immediately</u>?

Have your partner continue to smell the wintergreen oil until he is no longer able to detect the smell. Record his adaptation time on the data sheet. Then have him open his other nostril.

<u>9. Can your partner smell the wintergreen oil with both nostrils open</u>?

<u>10. Which cranial nerve carries smell sensations to the brain</u>?

<u>11. What is unusual about this sensory pathway</u>?

5.3. <u>Hearing</u>. Seat your partner with his eyes closed at the lab bench. Slowly bring a watch within hearing range from the front, side, back, and top of his head.

<u>12. From which direction is the sound localization best</u>?

Strike a tuning fork on the palm of your hand. Place the handle of the tuning fork on the frontal, parietal, temporal, and occipital skull bones.

<u>13. Which location gives the best bone conduction of sound waves</u>?

There are two tests that are commonly used to distinguish between conduction deafness and nerve deafness.

In the Weber test, conduction deafness is simulated by placing a ball of sterile cotton in one ear. Clean the handle of a tuning fork with an alcohol swab. Strike the tuning fork on the palm of your hand and place it on the bridge of your partner's nose. A person with normal hearing will hear the sound coming from a midline position. A person with conduction deafness (which we have created in one ear) will hear the sound better in the deaf ear. The reason for this is that sounds entering the normal ear will be partially masked by environmental noise to which the deaf ear is less sensitive.

In the Rinne test, conduction deafness is also simulated by placing a ball of cotton in one ear. Strike the tuning fork on the palm of your hand and place the handle on your partner's mastoid process. After the vibrations can no longer be

52

felt, place the fork near the deaf ear. A person with conduction deafness (which
we have created in one ear) will hear the fork longer by bone conduction. A
person with normal hearing will hear the sound in air after the bone conduction
is over. Try the Rinne test with the normal ear and with the deaf ear to demonstrate these features (Table 5.1).

Table 5.1 Tests for Hearing Loss

| CONDITION | FINDING | |
	WEBER TEST	RINNE TEST
No hearing loss	No lateralization	Sound heard longer by air conduction
Conduction deafness	Lateralization to poorer ear	Sound heard longer by bone conduction
Nerve deafness	Lateralization to better ear	Sound heard longer by air conduction

5.4. Vision. The blind spot (optic disc) is an area on the retina of each eye
where the optic nerve and blood vessels leave and enter the retina. As a result,
there are no rods or cones for light reception in these areas. Close your left eye
and focus your right eye on the "X" shown in Figure 5.1. The image of the "X" is
now focused on the light receptors in your right eye. Hold the "X" about 18 inches
from your eye and slowly bring it closer. At some point, the round dot will
disappear. At this distance, the dot is being focused on the blind spot of your
right eye.

Figure 5.1

14. What is the distance (cm) for your right eye?

Repeat the procedure for your left eye. Note that this time you must focus on
the dot initially and watch for the "X" to disappear.

15. What is the distance for your left eye?

Visual acuity refers to the ability of the eye to discriminate details. The
Snellen eye chart contains letters that should be seen clearly at specific dis-
tances by eyes with normal visual acuity. Thus the letters on line 1 of the chart
should be read easily at 200 feet, the letters on line 8 at 20 feet, and so on.

Visual acuity is stated as the ratio of two numbers. For example, if visual acuity is 20/30, it indicates that the eye must be 20 feet from what a normal eye sees at 30 feet.

To determine your visual acuity, stand 20 feet from the Snellen eye chart. Cover one eye with a 3 X 5 index card. Read the chart as far down as possible. Then repeat the procedure with the other eye.

16. What is the visual acuity for each of your eyes?

Obtain four sheets of white paper. On the first three, make a red, a green, and a yellow X, respectively, with colored masking tape. Stare at one of the X's for 30 seconds and then look at the blank sheet of white paper.

17. What do you see?

Repeat the procedure for the other two colors. A negative afterimage should be seen. This results from the adaptation or bleaching of a particular cone group in the retina. When you look at a white sheet of paper, these cones cannot respond, and thus this color is removed from the white spectrum. However, the unadapted cones are still sensitive to all colors of light and can still respond.

Color blindness is a sex-linked genetic abnormality. The most common type is red-green color blindness in which the person lacks either the red or green cone pigments in the retina. If the red cones are missing, the red wavelengths of light will be perceived as green. If the green cones are missing, only red wavelengths of light will be perceived.

The Ishihara charts are the most widely used in the testing of color blindness. The test consists of a collection of colored dots arranged in plates so that a person with normal color vision sees a number or pattern and a color-blind person sees a different number or pattern. Hold the charts about 30 inches from your eyes and try to read each plate. Compare your answers with those given in the front of the Ishihara test book.

Finally, have your partner hold a large test tube upright at eye level about 5 feet away from you. Close your left eye. Hold a pencil or pen vertically in your hand and approach the test tube. Try to place the pencil or pen in the test tube. Then try again with both eyes open.

18. Can you do it with only one eye open?

19. How does binocular vision help in depth perception?

DATA SHEET FOR EXPERIMENT 5: HUMAN SPECIAL SENSES

Name _____ Lab Section _____ Date _____

5.1. Gustation

1. Is the sucrose tasted immediately? _____

    Why? _____

Substance	Tip	Sides	Back
Sweet			
Sour			
Salty			
Bitter			

2. Which cranial nerves carry taste sensations to the brain? _____

    _____

3. Which sucrose solution was closest to the taste threshold for your partner?

    _____

4. Can your partner identify the food cube?_____

5. Can your partner identify the food cube after chewing?

    _____

6. Can your partner identify the food cube with open nostrils?

    _____

7. Can you taste PTC? _____

5.2. Olfaction

8. Is the smell detected immediately? _____

    Adaptation time _____ sec

9. Can your partner smell the wintergreen oil with both nostrils open?

    _____

10. Which cranial nerve carries smell sensations to the brain?

    _____

11. What is unusual about this sensory pathway? _____

_____

5.3. <u>Hearing</u>

12. From which direction is sound localization best?

_____

13. Which location gives the best bone conduction of sound waves?

_____

5.4. <u>Vision</u>

14. What is the distance (cm) for your right eye?

_____

15. What is the distance for your left eye? _____

16. What is the visual acuity for each of your eyes?

Right eye _____

Left eye _____

17. What do you see?

Red _____

Green _____

Yellow _____

18. Can you do it with only one eye open? _____

19. How does binocular vision help in depth perception?

_____

_____

EXPERIMENT 6

FROG SKELETAL MUSCLE CONTRACTION

Behavioral Objectives

The student should be able to:

1. State the characteristics of all contractile systems.
2. State the most studied skeletal and cardiac muscles.
3. Determine the twitch threshold of a skeletal muscle.
4. Construct a strength-duration curve from sample data.
5. Demonstrate summation, unfused tetanus, and fused tetanus.
6. Demonstrate muscle fatigue.

Materials

Large frogs
Wax dissecting pans
Dissecting pins
Fine thread
Dissecting instruments
Frog Ringer solution
Stimulator
Physiograph with myograph (or kymograph)
Myograph tension adjuster (or muscle lever)
Pin electrodes

Introduction

Muscle may be classified according to histological appearance (smooth, striated),
mode of central nervous system control (voluntary, involuntary), or by structures
associated with the muscle (skeletal, intestinal). Every muscle contractile system
studied thus far has been found to contain *actomyosin*. Actomyosin is composed of
the proteins actin, myosin, troponin, and tropomyosin. Thus the thing that
different types of muscle have in common is their contractile proteins.

Most of our knowledge of muscle has come from the study of individual muscle
preparations. The *frog sartorius muscle* has been used to gather data about muscle
energetics and mechanics, as well as neuromuscular transmission. In fact, the
frog sartorius muscle has the distinction of being the most studied skeletal
muscle. The *rabbit psoas muscle* has been used for studies of the biochemistry of
muscle contraction. Reptilian and insect skeletal muscles have been used for

57

electron microscope studies of muscle fine structure. Cat and rat uterine smooth muscles have been used in studies of smooth muscle. In addition, the *rabbit papillary muscle* is perhaps one of the most studied pieces of cardiac muscle.

The *frog gastrocnemius muscle*, the preparation used most often in the classroom, is rarely used for muscle research. The gastrocnemius muscle is thick and sturdy (hence its value in the classroom) but is difficult to see through, contains a mixture of fiber types, and is difficult to perfuse with drugs (hence making the interpretation of results difficult).

Experimental Procedure

In this experiment, the mechanical properties of the frog gastrocnemius muscle will be examined. Kill a frog by double pithing it, and cut the skin transversely around its chest just behind the forelegs. Strip the frog of its skin by gripping the skin with a pair of heavy forceps and pulling the skin down to the frog's feet. Try not to allow the outside surface of the skin to come in contact with the underlying muscle; some of the skin secretions are toxic to excitable cells.

Identify the gastrocnemius muscle, as shown in Figure 6.1 a, b, c, and d. Tie a piece of thread around the Achilles tendon below the thickening of the heel. Then cut the tendon below where the thread is tied. Free the muscle from its attachments to the tibia, but do not cut the muscle away from the knee. Cut off the tibia and other muscles below the knee joint. Clear away muscles in the area of the femur, and cut the femur in the middle of the thigh. Place the femur stump in the clamp as shown in Figure 6.2 a and b.

During the dissection and throughout the experiment, keep the muscle moist with frog Ringer solution. Hang the gastrocnemius muscle vertically between the force transducer and femur clamp as shown in Figure 6.2 a and b. Keep the thread between the tendon and the transducer as short as possible, and make sure that the muscle is taut. It is important, however, not to stretch the muscle beyond "rest length"—the normal relaxed length found *in vivo*—or the muscle will begin to deteriorate rapidly.

Kymograph Procedure

Position the writing tips of the signal marker and the muscle lever so that they are writing on the same vertical line. Attach the thread tied to the Achilles tendon to the muscle lever of the kymograph. Then adjust the height of the femur clamp so that the muscle is at rest length and the thread is taut. Attach the pin electrodes to the muscle by carefully piercing each end of the muscle with an electrode (Figure 6.2 a and b).

6.1. **Twitch Threshold.** Place a 5-g weight in the pan attached to the muscle lever and adjust the afterloading screw so that the muscle is not stretched. Set the kymograph speed knob to position 3.
Initial stimulator settings:

Voltage set at 0 V
Duration set at 100 msec
Mode switch set to deliver single stimuli

Turn on the recorder and deliver a single stimulus to the muscle by pressing the mode switch. Increase the strength of the stimulus voltage a small amount and repeat until the muscle shows a small contraction (Figure 6.3).

(a)

Figure 6.1 Dissection of frog gastrocnemius muscle. (a) Frog with dorsal surface
facing upward.

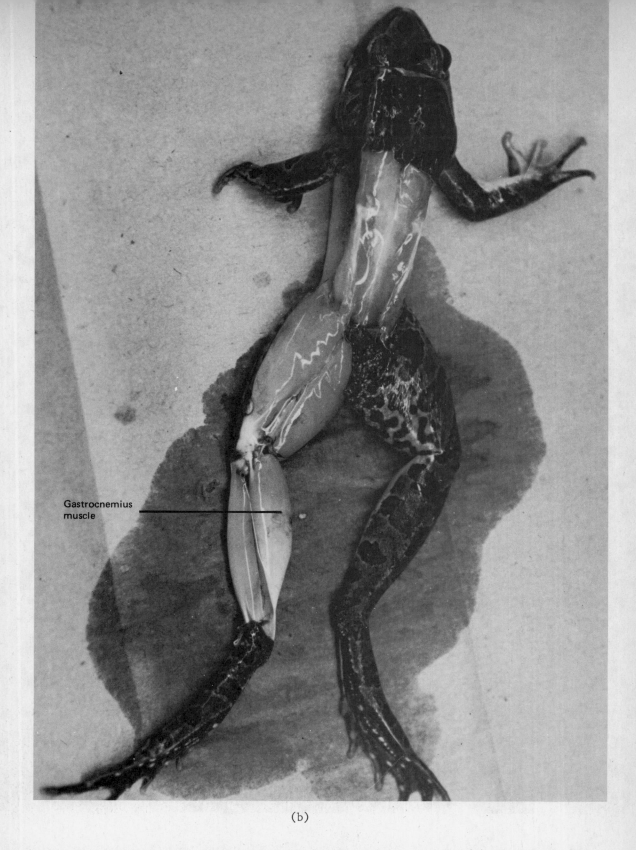

Gastrocnemius
muscle

(b)

Figure 6.1 (Continued) Dissection of frog gastrocnemius muscle. (b) Location of
    gastrocnemius muscle.

60

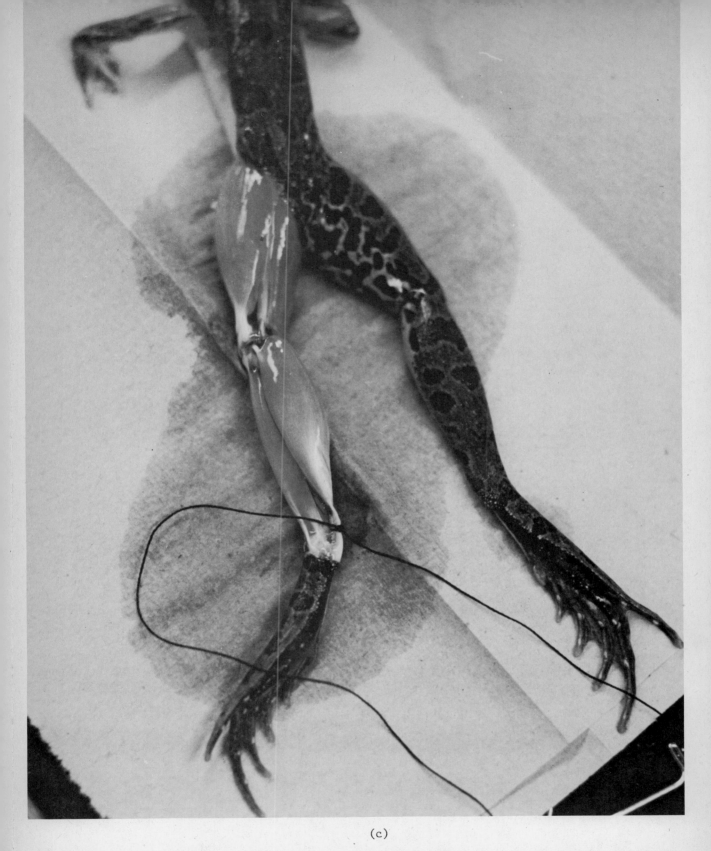

(c)

Figure 6.1 Dissection of frog gastrocnemius muscle. (c) Achilles tendon tied with
thread.

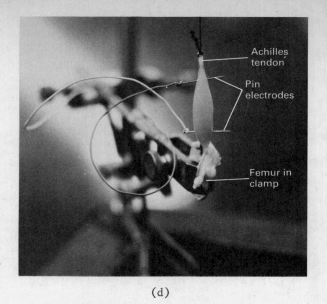

(d)

Figure 6.1 (Continued) Dissection of frog gastrocnemius muscle. (d) Isolated gastrocnemius muscle.

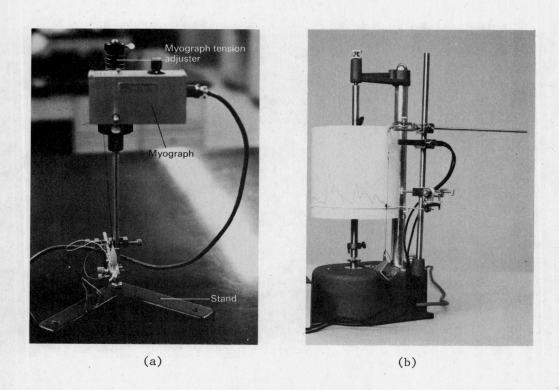

(a)   (b)

Figure 6.2 Mounting of gastrocnemius muscle. (a) Physiograph setup (b) Kymograph setup. (Photograph courtesy of Phipps and Bird, Inc.)

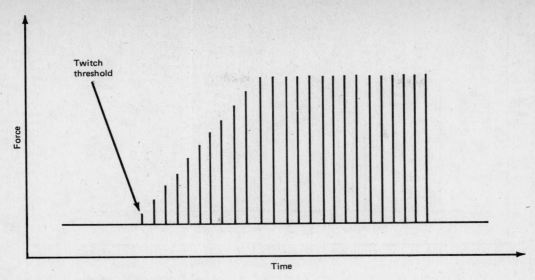

Figure 6.3 The twitch threshold.

1. What is the threshold voltage?

    Continue increasing the stimulus voltage until no further increase in contraction is noted.

2. What is the maximum stimulus voltage?

    Repeat the above procedure using stimulus durations of 25, 50, and 120 msec. In each case, find the stimulus strength (V) that gives maximal twitch tension. Draw a strength-duration curve by plotting stimulus strength versus stimulus duration.

3. What can you conclude from this relationship?

    6.2. Length-Tension Relationship. If a muscle is stimulated after its length has been changed from rest (normal) length, the amount of tension that it can develop is less than maximal. Our setup allows us to study the effect of stretching the muscle to lengths greater than rest length (Figure 6.4).
    Add a 5-g weight to the pan and record a single twitch. Use a stimulus strength and duration that gave a maximal response in the previous section of the experiment. If no decrease in twitch amplitude is seen, add another 5-g weight to the pan and obtain another twitch.

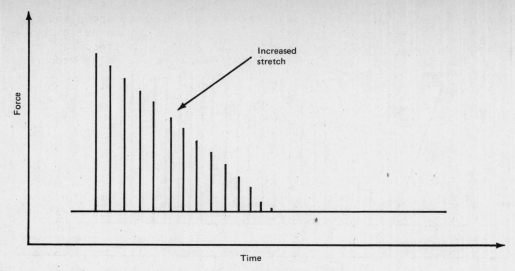

Figure 6.4   The effect of stretch on force development.

4. **Does the amplitude of the twitch fall as more and more stretching is applied to the muscle?   Why?**

    6.3.   **Summation and Tetanus.**   When two stimuli are delivered to the muscle close together in time, the second stimulus will produce a contraction that adds onto the first contraction.   This is called *summation*.   As the stimulation frequency is increased, individual twitches continue to sum together to produce a jagged contraction called *unfused tetanus*.   When the stimulation frequency is very rapid, a smooth contraction called *fused tetanus* is produced in which no relaxation occurs between individually summed twitches (Figure 6.5).

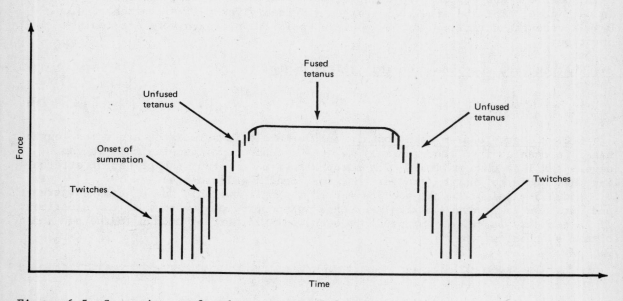

Figure 6.5   Summation, unfused tetanus, and fused tetanus.

64

Adjust the speed control knob on the kymograph to maximum speed and turn on the recorder. Stimulate the muscle with a voltage and duration pulse that gave maximal tension previously. Gradually increase the stimulus frequency until summation and tetanus occur. Then gradually decrease the stimulus frequency to its original level.

5. <u>Why is tetanic tension greater than twitch tension</u>?

    <u>6.4. Fatigue</u>. Fatigue is caused by the depletion of ATP or oxygen within a muscle. The onset of fatigue can be hastened by the accumulation of waste products from an earlier fatigue. Fatigue can be seen in our preparation by a gradual loss of tension over time in the sustained presence of tetanizing stimuli (Figure 6.6).
    Adjust the afterloading screw so that the muscle carries no weight but the thread is taut. Add a 10-g weight to the pan. Set the kymograph speed to its lowest value and turn on the recorder. Stimulate the muscle with stimuli that give maximal tension and use a low tetanizing frequency. Continue stimulation until the contraction amplitude decreases to one-half of its initial amplitude.

6. <u>How long does fatigue to one-half maximal take</u>?

Wait 5 minutes and apply a single, supramaximal stimulus to the muscle.

7. <u>What is the amplitude of the twitch</u>?

Repeat the fatigue a second time and, after waiting 5 minutes, elicit another twitch from the muscle.

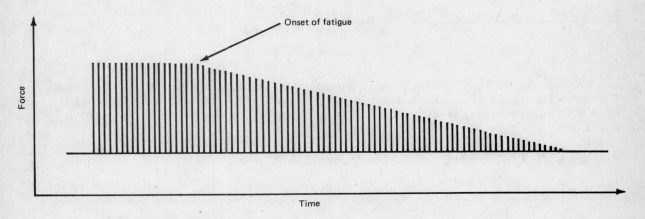

Figure 6.6   Skeletal muscle fatigue.

8. What is the second fatigue time?

9. What is the amplitude of the twitch?

10. Why do the time to fatigue and the twitch amplitude change?

Physiograph Procedure

Attach the thread tied to the Achilles tendon to the hook of the force trans-ducer. Then adjust the myograph tension adjuster so that the muscle is at rest length and the thread is taut. Attach the pin electrodes to the muscle by care-fully piercing each end of the muscle with an electrode.

6.1. Twitch Threshold. Run the physiograph at a slow paper speed.

Initial stimulator settings:

Voltage set at 0 V
Duration set at 100 msec
Mode switch set to deliver single stimuli

Start the paper moving and deliver a single stimulus to the muscle by pressing the mode switch. Increase the strength of the stimulus voltage a small amount and repeat until the muscle shows a small contraction (Figure 6.3).

1. What is the threshold voltage?

Continue increasing the stimulus voltage until no further increase in contrac-tion is noted.

2. What is the maximal stimulus voltage?

Repeat the above procedure using stimulus durations of 25, 50, and 120 msec. In each case, find the stimulus strength (V) that gives maximal twitch tension. Draw a strength-duration curve by plotting stimulus strength versus stimulus duration.

3. What can you conclude from this relationship?

6.2. <u>Length-Tension Relationship</u>. If a muscle is stimulated after its length has been changed from rest (normal) length, the amount of tension that it can develop is less than maximal. Our setup permits us to study the effect of stretching the muscle to lengths greater than rest length (Figure 6.4).

Turn the myograph tension adjuster so as to stretch the muscle a small amount. Stimulate the muscle with a stimulus that gave a maximal response in the previous section of the experiment. If no decrease in twitch amplitude is seen, stretch the muscle a little more.

<u>4.</u> <u>Does the amplitude of the twitch fall as more and more stretching is applied to the muscle? Why?</u>

6.3. <u>Summation and Tetanus</u>. When two stimuli are delivered to the muscle close together in time, the second stimulus will produce a contraction that adds onto the first contraction. This is called *summation*. As the stimulation frequency is increased, individual twitches continue to sum together to produce a jagged contraction called *unfused tetanus*. When the stimulation frequency is very rapid, a smooth contraction called *fused tetanus* is produced in which no relaxation occurs between individually summed twitches (Figure 6.5).

Adjust the paper speed on the physiograph to 1 cm/sec and start the paper moving. Stimulate the muscle with a voltage and duration pulse that gave maximal tension previously. Gradually increase the stimulus frequency until summation and tetanus occur. Then gradually decrease the stimulus frequency to its original level.

<u>5.</u> <u>Why is tetanic tension greater than twitch tension?</u>

6.4. <u>Fatigue</u>. Fatigue is caused by the depletion of ATP or oxygen within a muscle. The onset of fatigue can be hastened by the accumulation of waste products from an earlier fatigue. Fatigue can be seen in our preparation by a gradual loss of tension over time in the sustained presence of tetanizing stimuli (Figure 6.6).

Set the physiograph paper speed to its lowest value and start the paper moving. Stimulate the muscle with stimuli that give maximal tension and use a low tetanizing frequency. Continue stimulation until the contraction amplitude decreases to one-half of its initial amplitude.

<u>6.</u> <u>How long does fatigue to one-half maximal take?</u>

Wait 5 minutes and apply a single, supramaximal stimulus to the muscle.

<u>7.</u> <u>What is the amplitude of the twitch?</u>

Repeat the fatigue a second time and, after waiting 5 minutes, elicit another twitch from the muscle.

67

8. What is the second fatigue time?

9. What is the amplitude of the twitch?

10. Why do the time to fatigue and the twitch amplitude change?

Name _____ Lab Section _____ Date _____

6.1. Twitch Threshold

1. What is the threshold voltage? _____

2. What is the maximal stimulus voltage? _____

Stimulus Duration (msec)	Stimulus Strength (V)
25	
50	
100	
120	

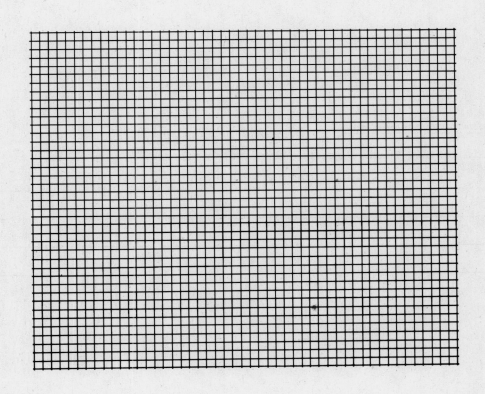

3. What can you conclude from this relationship?

_____

_____

_____

6.2. Length-Tension Relationship

4. Does the amplitude of the twitch tension fall as more and more stretching is
applied to the muscle? Why? _____

_____

_____

Sample Record—Length-Tension Relationship

6.3. Summation and Tetanus

5. Why is tetanic tension greater than twitch tension?

_____

_____

_____

_____

6.4. Fatigue

6. How long does fatigue to one-half maximal take? _____

7. What is the amplitude of the twitch? _____

8. What is the second fatigue time? _____

9. What is the amplitude of the twitch? _____

10. Why do the time to fatigue and the twitch amplitude change? _____

_____

_____

_____

Sample Record—Fatigue

EXPERIMENT 7

HUMAN RESPIRATION

Behavioral Objectives

The student should be able to:

1. Describe percussion, palpation, and ausculation.
2. Define eupnea, dyspnea, hypercapnia, apnea, and orthopnia.
3. Construct a spirogram.
4. Measure a peripheral pulse contour.
5. State the significance of the dicrotic notch.
6. Describe how respiration and pulse rates vary with body position.
7. Measure lung volumes and capacities by using a spirometer.
8. Demonstrate the effects of hyperventilation and rebreathing on respiration and pulse rates.
9. Demonstrate the effects of reading aloud, reading silently, and adding on respiration and pulse rates.

Materials

Dry spirometers with mouthpieces
Physiograph with bellows pneumograph (or kymograph with Marey tambour)
Stethoscopes
Alcohol swabs
Lunch bags
Meter sticks

Introduction

Oxygen is one of the basic needs of most living things for carrying out cellular metabolism. In lower animals, oxygen is obtained directly from the environment by simple diffusion. In higher animals, special respiratory systems have developed which provide a moist environment for the diffusion of oxygen and carbon dioxide in and out of the blood.

The human respiratory system is made up of a series of passageways that conduct air from the environment to the lungs. As air is passing through these passageways, *respiratory sounds* can be heard on the surface of the body. Air passing through small bifurcations produces turbulent flow which gives rise to inspiratory sounds. Air passing out of the lungs does not encounter sharp bifurcations and thus produces less noise than inspired air. Since alveoli, the sites of gas exchange within the lungs, tend to damp vibrations, a diseased lung will produce louder noises than a healthy lung because nonfunctional alveoli will not dampen the sounds and they will be transmitted to the chest wall.

Percussion refers to the tapping of the chest to determine the position, size, or consistency of underlying tissue. The pitch of the sound produced by percussion is determined by the ratio of air-filled tissue to solid tissue. Air-filled tissue produces a low-pitched, resonant sound; solid tissue produces a high-pitched sound.

Palpation refers to the external examination of the body by feeling with the hands. Vocal sounds (*fremitus*) cause vibrations that can be palpated (tactile fremitus) and heard (auditory fremitus).

Ausculation refers to listening to body sounds for the purpose of detecting abnormal conditions. Normal breathing sounds can be heard over the entire surface of the lungs except for the apex of the right lung where a mixture of sounds occurs. Normal sounds are more easily heard during inspiration for the reasons mentioned above.

The measurement of respiratory volumes and capacities is important in estimating respiratory function. In various disease states, the normal lung volumes and capacities are reduced. If fluid accumulates in areas normally filled with air, a decrease in the ability of the lungs to carry out gas exchange will occur. Conditions such as emphysema, pneumonia, asthma, tuberculosis, and cigarette smoking all decrease certain of the lung volumes and capacities.

Experimental Procedure

7.1. Respiratory Sounds. Tap your partner's chest in several areas on both the front and rear of the chest.

1. Are all of the sounds heard alike in response to this percussion? Do they differ in pitch or duration?

Make a low vocal sound. Have your partner palpate your chest in various areas during the sound.

2. Where are the vibrations most noticeable?

3. Which sex would have greater tactile fremitus? Why?

Place a stethoscope over the upper right chest area of your partner. Clean the earpieces of the stethoscope with an alcohol swab before placing the earpieces in your ears. Listen for breathing sounds in your partner's chest.

4. Where are the sounds most prominent?

5. <u>Describe the differences in the sounds heard</u>.

<u>7.2</u>. <u>Respiratory Volumes and Capacities</u>. The *spirometer* is an instrument that is used to measure respiratory volumes. Spirometers may be of two types, wet or dry. We will use a simple, hand-held dry type.

Clean off the end of the spirometer stem with an alcohol swab and insert a disposable mouthpiece on the stem. The dry spirometer is capable of measuring air volumes only when air is exhaled into its mouthpiece. As a result, it is important to pinch your nostrils closed when exhaling into the spirometer to prevent air from being lost through the nostrils. Hold the spirometer so that the dial faces upward. The dial reads in cubic centimeters of air; set the dial to read 1000 cm^3. This initial reading will then be subtracted from each final reading to give the desired lung volume or capacity.

The *tidal volume* (TV) is the total amount of air that moves in and out of the lungs during one inspiration and one expiration. The tidal volume depends on age, sex, and physical size, and is about 500 cm^3 in an average adult. After you have been breathing normally, expire into the spirometer three times while inhaling through the nose between breaths. Subtract 1000 cm^3 from the total volume recorded and divide this value by three to obtain your tidal volume. Fill in the appropriate spaces on the data sheet.

The *expiratory reserve volume* (ERV) is the amount of air that can be forcibly exhaled beyond the tidal volume. The ERV normally amounts to about 1100 cm^3. Reset the spirometer dial to 1000 cm^3. After you have been breathing normally, expel all of the air that you can into the spirometer. Subtract 1000 cm^3 and your tidal volume from the reading. Repeat this procedure three times and obtain your mean ERV.

The *vital capacity* (VC) is the sum of the tidal volume, inspiratory reserve volume, and expiratory reserve volume. The average vital capacity for men and women is about 4500 cm^3. However, age, sex, and physical size affect the vital capacity extensively. As a result, a natural variation of ±20 percent is considered normal. Set the spirometer dial on 1000 cm^3. After taking three *deep* breaths, take one final deep breath and exhale as much air as possible into the spirometer. A slow, forced exhalation works best. Then subtract 1000 cm^3 from the final reading to obtain your VC. Repeat this procedure three times to get your mean VC. Compare your VC with those in Tables 7.1 and 7.2.

The *inspiratory reserve volume* (IRV) is the amount of air that can be forcibly inhaled with a deep breath. It is usually about 2800 cm^3. Calculate your IRV by subtracting the sum of your TV and ERV from your VC as measured above.

The *residual volume* (RV) is the volume of air in the lungs that cannot be forcibly exhaled. The residual volume normally amounts to about 1200 cm^3, but it cannot be measured by simple spirometric methods.

<u>7.3</u>. <u>Measurement of Breathing Patterns</u>. Seat your partner on the lab bench so that he or she cannot see the recording apparatus.

<u>Physiograph Method</u>

Attach a bellows pneumograph snugly to your partner's chest with the leather strap provided. During the attachment, make sure the wingnut on the transducer is open. After tightening the strap, compress the rubber bellows slightly and at the same time tighten the wingnut. Then connect the transducer to the transducer coupler of the physiograph by means of a transducer cable.

Table 7.1 Predicted Vital Capacities for Males

HEIGHT IN CENTIMETERS AND INCHES

AGE	CM. 152 / IN. 59.8	154 / 60.6	156 / 61.4	158 / 62.2	160 / 63.0	162 / 63.7	164 / 64.6	166 / 65.4	168 / 66.1	170 / 66.9	172 / 67.7	174 / 68.5	176 / 69.3	178 / 70.1	180 / 70.9	182 / 71.7	184 / 72.4	186 / 73.2	188 / 74.0
16	3920	3975	4025	4075	4130	4180	4230	4285	4335	4385	4440	4490	4540	4590	4645	4695	4745	4800	4850
18	3890	3940	3995	4045	4095	4145	4200	4250	4300	4350	4405	4455	4505	4555	4610	4660	4710	4760	4815
20	3860	3910	3960	4015	4065	4115	4165	4215	4265	4320	4370	4420	4470	4520	4570	4625	4675	4725	4775
22	3830	3880	3930	3980	4030	4080	4135	4185	4235	4285	4335	4385	4435	4485	4535	4585	4635	4685	4735
24	3785	3835	3885	3935	3985	4035	4085	4135	4185	4235	4285	4330	4380	4430	4480	4530	4580	4630	4680
26	3755	3805	3855	3905	3955	4000	4050	4100	4150	4200	4250	4300	4350	4395	4445	4495	4545	4595	4645
28	3725	3775	3820	3870	3920	3970	4020	4070	4115	4165	4215	4265	4310	4360	4410	4460	4510	4555	4605
30	3695	3740	3790	3840	3890	3935	3985	4035	4080	4130	4180	4230	4275	4325	4375	4425	4470	4520	4570
32	3665	3710	3760	3810	3855	3905	3950	4000	4050	4095	4145	4195	4240	4290	4340	4385	4435	4485	4530
34	3620	3665	3715	3760	3810	3855	3905	3950	4000	4045	4095	4140	4190	4225	4285	4330	4380	4425	4475
36	3585	3635	3680	3730	3775	3825	3870	3920	3965	4010	4060	4105	4155	4200	4250	4295	4340	4390	4435
38	3555	3605	3650	3695	3745	3790	3840	3885	3930	3980	4025	4070	4120	4165	4210	4260	4305	4350	4400
40	3525	3575	3620	3665	3710	3760	3805	3850	3900	3945	3990	4035	4085	4130	4175	4220	4270	4315	4360
42	3495	3540	3590	3635	3680	3725	3770	3820	3865	3910	3955	4000	4050	4095	4140	4185	4230	4280	4325
44	3450	3495	3540	3585	3630	3675	3725	3770	3815	3860	3905	3950	3995	4040	4085	4130	4175	4220	4270
46	3420	3465	3510	3555	3600	3645	3690	3735	3780	3825	3870	3915	3960	4005	4050	4095	4140	4185	4230
48	3390	3435	3480	3525	3570	3615	3655	3700	3745	3790	3835	3880	3925	3970	4015	4060	4105	4150	4190
50	3345	3390	3430	3475	3520	3565	3610	3650	3695	3740	3785	3830	3870	3915	3960	4005	4050	4090	4135
52	3315	3353	3400	3445	3490	3530	3575	3620	3660	3705	3750	3795	3835	3880	3925	3970	4010	4055	4100
54	3285	3325	3370	3415	3455	3500	3540	3585	3630	3670	3715	3760	3800	3845	3890	3930	3975	4020	4060
56	3255	3295	3340	3380	3425	3465	3510	3550	3595	3640	3680	3725	3765	3810	3850	3895	3940	3980	4025
58	3210	3250	3290	3335	3375	3420	3460	3500	3545	3585	3630	3670	3715	3755	3800	3840	3880	3925	3965
60	3175	3220	3260	3300	3345	3385	3430	3470	3500	3555	3595	3635	3680	3720	3760	3805	3845	3885	3930
62	3150	3190	3230	3270	3310	3350	3390	3440	3480	3520	3560	3600	3640	3680	3730	3770	3810	3850	3890
64	3120	3160	3200	3240	3280	3320	3360	3400	3440	3490	3530	3570	3610	3650	3690	3730	3770	3810	3850
66	3070	3110	3150	3190	3230	3270	3310	3350	3390	3430	3470	3510	3550	3600	3640	3680	3720	3760	3800
68	3040	3080	3120	3160	3200	3240	3280	3320	3360	3400	3440	3480	3520	3560	3600	3640	3680	3720	3760
70	3010	3050	3090	3130	3170	3210	3250	3290	3330	3370	3410	3450	3480	3520	3560	3600	3640	3680	3720
72	2980	3020	3060	3100	3140	3180	3210	3250	3290	3330	3370	3410	3450	3490	3530	3570	3610	3650	3680
74	2930	2970	3010	3050	3090	3130	3170	3200	3240	3280	3320	3360	3400	3440	3470	3510	3550	3590	3630

Source: Table 7.1 is taken from Tables D and E of Gaensler and Wright: "Evaluation of Respiratory Impairment," *Arch. Environ. Health, 12*: 184-185, 1966, reprinted with permission of Heldref Publications, 4000 Albemarle St., N.W., Washington, D.C.

Table 7.2 Predicted Vital Capacities for Females

AGE	HEIGHT IN CENTIMETERS AND INCHES																		
CM. IN.	152 59.8	154 60.6	156 61.4	158 62.2	160 63.0	162 63.7	164 64.6	166 65.4	168 66.1	170 66.9	172 67.7	174 68.5	176 69.3	178 70.1	180 70.9	182 71.7	184 72.4	186 73.2	188 74.0
16	3070	3110	3150	3190	3230	3270	3310	3350	3390	3430	3470	3510	3550	3590	3630	3670	3715	3755	3800
17	3065	3095	3135	3175	3215	3255	3295	3335	3375	3415	3455	3495	3535	3575	3615	3655	3695	3740	3780
18	3010	3080	3120	3160	3200	3240	3280	3320	3360	3400	3440	3480	3520	3560	3600	3640	3680	3720	3760
20	3010	3050	3090	3130	3170	3210	3250	3290	3330	3370	3410	3450	3490	3525	3565	3605	3645	3695	3720
22	2980	3020	3060	3095	3135	3175	3215	3255	3290	3330	3370	3410	3450	3490	3530	3570	3610	3650	3685
24	2950	2985	3025	3065	3100	3140	3180	3220	3260	3300	3335	3375	3415	3455	3490	3530	3570	3610	3650
26	2920	2960	3000	3035	3070	3110	3150	3190	3230	3265	3300	3340	3380	3420	3455	3495	3530	3570	3610
28	2890	2930	2965	3000	3040	3070	3115	3155	3190	3230	3270	3305	3345	3380	3420	3460	3495	3535	3570
30	2860	2895	2935	2970	3010	3045	3085	3120	3160	3195	3235	3270	3310	3345	3385	3420	3460	3495	3535
32	2825	2865	2900	2940	2975	3015	3050	3090	3125	3160	3200	3235	3275	3310	3350	3385	3425	3460	3495
34	2795	2835	2870	2910	2945	2980	3020	3055	3090	3130	3165	3200	3240	3275	3310	3350	3385	3425	3460
36	2765	2805	2840	2875	2910	2950	2985	3020	3060	3095	3130	3165	3205	3240	3275	3310	3350	3385	3420
38	2735	2770	2810	2845	2880	2915	2950	2990	3025	3060	3095	3130	3170	3205	3240	3275	3310	3350	3385
40	2705	2740	2775	2810	2850	2885	2920	2955	2990	3025	3060	3095	3135	3170	3205	3240	3275	3310	3345
42	2675	2710	2745	2780	2815	2850	2885	2920	2955	2990	3025	3060	3100	3135	3170	3205	3240	3275	3310
44	2645	2680	2715	2750	2785	2820	2855	2890	2925	2960	2995	3030	3060	3095	3135	3165	3200	3235	3270
46	2615	2650	2685	2715	2750	2785	2820	2855	2890	2925	2960	2995	3030	3060	3095	3130	3165	3200	3235
48	2585	2620	2650	2685	2715	2750	2785	2820	2855	2890	2925	2960	2995	3030	3060	3095	3130	3160	3195
50	2555	2590	2625	2655	2690	2720	2755	2785	2820	2855	2890	2925	2955	2990	3025	3060	3090	3125	3155
52	2525	2555	2590	2625	2655	2690	2720	2755	2790	2820	2855	2890	2925	2955	2990	3020	3055	3090	3125
54	2495	2530	2560	2590	2625	2655	2690	2720	2755	2790	2820	2855	2885	2920	2950	2985	3020	3050	3085
56	2460	2495	2525	2560	2590	2625	2655	2690	2720	2755	2790	2820	2855	2885	2920	2950	2980	3015	3045
58	2430	2460	2495	2525	2560	2590	2625	2655	2690	2720	2750	2785	2815	2850	2880	2920	2945	2975	3010
60	2400	2430	2460	2495	2525	2560	2590	2625	2655	2685	2720	2750	2780	2810	2845	2875	2915	2940	2970
62	2370	2405	2435	2465	2495	2525	2560	2590	2620	2655	2685	2715	2745	2775	2810	2840	2870	2900	2935
64	2340	2370	2400	2430	2465	2495	2525	2555	2585	2620	2650	2680	2710	2740	2770	2805	2835	2865	2895
66	2310	2340	2370	2400	2430	2460	2495	2525	2555	2585	2615	2645	2675	2705	2735	2765	2800	2825	2860
68	2280	2310	2340	2370	2400	2430	2460	2490	2520	2550	2580	2610	2640	2670	2700	2730	2760	2795	2820
70	2250	2280	2310	2340	2370	2400	2425	2455	2485	2515	2545	2575	2605	2635	2665	2695	2725	2755	2780
72	2220	2250	2280	2310	2335	2365	2395	2425	2455	2480	2510	2540	2570	2600	2630	2660	2685	2715	2745
74	2190	2220	2245	2275	2305	2335	2360	2390	2420	2450	2475	2505	2535	2565	2590	2620	2650	2680	2710

Source: Table 7.2 is taken from Tables D and E of Gaensler and Wright: "Evaluation of Respiratory Impairment," *Arch. Environ. Health, 12:* 184–185, 1966, reprinted with permission of Heldref Publications, 4000 Albemarle St., N.W., Washington, D.C.

Initial settings of channel amplifier:

Polarity—set to +
Filter—set to 10 kHz
Sensitivity—set to 100 mV/cm

The bellows pneumograph and physiograph yield data similar to those obtained with a *wet* spirometer. Run the physiograph at a paper speed of 0.1 cm/sec so that inspirations produce upward pen deflections and expirations produce downward pen deflections.

Kymograph Method

Attach a pneumograph to your partner's chest with the leather strap provided. Clamp the metal tube of a Marey tambour to a ringstand so that the diaphragm of the tambour is facing upward. Then attach the yoke to the tambour. Attach the tambour pen to the yoke so that it extends near the kymograph paper.

Connect one end of the pneumograph to a Y tube. Connect the other end of the Y tube to the Marey tambour. At the end of an inspiration, close the open end of the Y tube with a pinch clamp.

Initial settings:

Drum speed—set to speed 5

With your partner seated, record the resting respiration pattern for 1 minute. There are several terms that are used to describe breathing patterns:

Eupnea—normal, quiet breathing
Dyspnea—labored, painful breathing
Hypercapnia—rapid breathing caused by an increase in pCO_2
Apnea—temporary cessation of breathing
Orthopnia—inability to breathe in a horizontal position

Construct a spirogram similar to the one shown in Figure 7.1. This spirogram can be calibrated by making use of the tidal volume measured earlier with the dry spirometer. The minute respiratory volume is the amount of air that passes in and out of the lungs in 1 minute. Calculate the minute volume for your partner by multiplying the respiration rate by the tidal volume.

6. What is the minute respiratory volume for your partner?

7.4. Measurement of Peripheral Pulse Contour. The *peripheral pulse* is caused by the transient expansion and recoil of arterial walls. The appearance of a normal pulse wave may vary with the position and measuring device used. We will use a photoelectric pulse transducer to record the pulse waveform on the physiograph.

Connect the photoelectric pulse transducer to the input of the transducer coupler of the physiograph. The small red light in the detecting head should glow. Attach the detecting head to your partner's index finger with the Velcro light shield provided. Be careful to attach the transducer firmly but not so tight as to occlude blood flow. Run the paper at a speed of 2.5 cm/sec and obtain a recording of the pulse wave contour. Compare your recording with Figure 7.2, and identify the dicrotic notch. The *dicrotic notch* is caused by closure of the aortic semilunar valve. The height of the contour varies according to the degree of relaxation of the subject and the rate varies according to the heart rate.

78

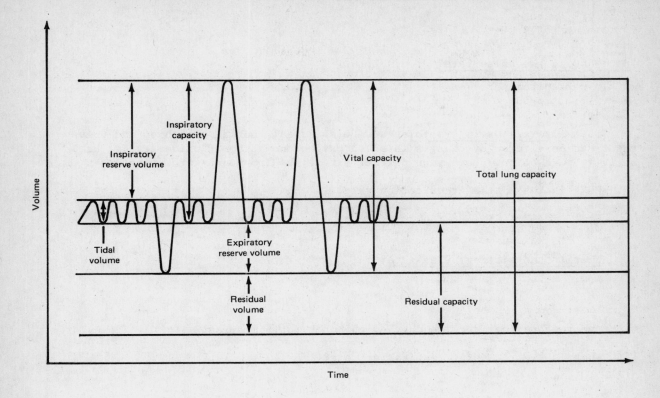

Figure 7.1   A typical spirogram.

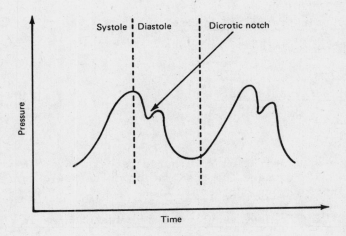

Figure 7.2   The peripheral pulse contour.

79

7. What is the height (cm) of the pulse contour?

8. What is the pulse rate?

   7.5. Alterations in Breathing and Pulse Rates. Normal breathing and pulse
rates can be altered by changes in blood oxygen and carbon dioxide levels, brain
activity, and so on. Have your partner sit at the lab bench for 2 minutes.

9. What is the sitting respiration rate?

10. What is the sitting pulse rate?

   Have your partner recline on the lab bench for 5 minutes.

11. What is the reclining respiration rate?

12. What is the reclining pulse rate?

   Have your partner stand up for 5 minutes.

13. What is the standing respiration rate?

14. What is the standing pulse rate?

   The normal ratio of respiration rate:pulse rate is 14 to 16 breaths:70 to
72 beats, or about 1:5. Compare the respiration:pulse ratio for sitting, reclining, and standing for your partner.

15. Does the ratio vary or stay about the same with changes in posture?

Have your partner hyperventilate for 1 minute and then record the respiration and pulse rates.

16. What is the respiration rate after hyperventilation?

17. What is the pulse rate after hyperventilation?

18. Why does hyperventilation sometimes lead to dizziness?

Have your partner recover for a few minutes and then have him rebreathe and hyperventilate into a paper bag for 1 minute. Make sure that both his nose and mouth are covered. Then record his respiration and pulse rates.

19. What is the respiration rate after rebreathing?

20. What is the pulse rate after rebreathing?

Record the respiration rate and pulse rate while your partner reads aloud, reads silently, and adds a column of numbers.

21. What is the respiration rate when reading aloud?

22. What is the pulse rate when reading aloud?

23. What is the respiration rate when reading silently?

24. What is the pulse rate when reading silently?

25. <u>What is the respiration rate when adding?</u>

26. <u>What is the pulse rate when adding?</u>

Name _____ Lab Section _____ Date _____

7.1. Respiratory Sounds

1. Are all of the sounds heard alike in response to this percussion? _____

   Do they differ in pitch and duration? _____

2. Where are the vibrations most noticeable? _____

   _____

3. Which sex would have greater tactile fremitus? _____

   Why? _____

4. Where are the sounds most prominent? _____

5. Describe the differences in the sounds heard. _____

   _____

7.2. Respiratory Volumes and Capacities

Volume or Capacity	Trial			Average (ml)
	1	2	3	
Tidal Volume				
Exp. Res. Vol.				
Vital Capacity				
Insp. Res. Vol.				

7.3. Measurement of Breathing Patterns

Sample Spirogram:

6. What is the minute respiratory volume for your partner?

_____

7.4. Measurement of Peripheral Pulse Contour

Sample Peripheral Pulse Contour:

7. What is the height (cm) of the pulse contour? _____

8. What is the pulse rate? _____

7.5. Alterations in Breathing and Pulse Rates

9.   What is the sitting respiration rate? _____

10.   What is the sitting pulse rate? _____

11.   What is the reclining respiration rate? _____

12.   What is the reclining pulse rate? _____

13.   What is the standing respiration rate? _____

14.   What is the standing pulse rate? _____

15.   Does the ratio vary or stay about the same with changes in posture? _____

16.   What is the respiration rate after hyperventilation? _____

17.   What is the pulse rate after hyperventilation? _____

18.   Why does hyperventilation sometimes lead to dizziness? _____

_____

19.   What is the respiration rate after rebreathing? _____

20.   What is the pulse rate after rebreathing? _____

21.   What is the respiration rate when reading aloud? _____

22.   What is the pulse rate when reading aloud? _____

23.   What is the respiration rate when reading silently? _____

24.   What is the pulse rate when reading silently? _____

25.   What is the respiration rate when adding? _____

26.   What is the pulse rate when adding? _____

EXPERIMENT 8

FROG HEART PHYSIOLOGY

Behavioral Objectives

The student should be able to:

1. Distinguish between myogenic and neurogenic hearts.
2. Demonstrate the effects of temperature on heart rate and force of contraction.
3. Demonstrate the effects of acetylcholine, epinephrine, and atropine on heart rate and force of contraction.
4. Simulate heart block.

Materials

Large frogs
Wax dissecting pans
Dissecting pins
Fine thread
Dissecting instruments
Frog Ringer solution
Physiograph with myograph (or kymograph)
Myograph tension adjuster (or heart lever)
Small hooks
Ice buckets with ice
Hot water bath (37°C)
Thermometers (0 to 100°C)
Beakers (50 ml)
Acetylcholine Ringer
Epinephrine Ringer
Atropine Ringer
Stimulator

Introduction

Hearts can be classified on the basis of structure as chambered, tubular, vesicular, or ampular. All hearts have three common properties: (1) a *pacemaker system*, muscular or nervous, where excitation originates, (2) a *conducting system*, muscular or nervous, which transports excitation by action potentials, and (3) a *contractile system*, made up of proteins, which performs work.

87

It is convenient for us to classify hearts according to the origin of the excitation and the type of pacemaker in operation. In *myogenic hearts*, excitation spontaneously originates in modified muscle tissue and nerve activity modulates this spontaneous rhythmic beating. In *neurogenic hearts*, nerve ganglion cells rhythmically discharge to initiate the beating and the heart muscle itself is not spontaneously active. These two types of hearts may be distinguished from each other by:

1. The presence or absence of ganglion cells and the effects of removal of these cells on the heartbeat.
2. The detection of the pacemaker by local warming or cooling or by excision of different regions to determine if the pacemaker is nervous or muscular.
3. The determination of the point of origin of the electrocardiogram.
4. The shape and form of the electrocardiogram—spikes and oscillations occur in neurogenic hearts; large slow waves occur in noninnervated hearts.
5. The activity of the heart tissue under noninnervated conditions, for example, in embryonic tissue or in tissue culture.
6. The effects of various drugs:
   Acetylcholine—stimulates ganglionic transmission; inhibits myogenic hearts; no effect on noninnervated myogenic hearts.
   Epinephrine—stimulates myogenic hearts
   Atropine—blocks cholinergic receptors

Experimental Procedure

Double pith a frog and expose the heart as shown in Figure 8.1a, b, and c. Free the heart from surrounding tissue. Grasp the tip of the ventricle with a pair of fine forceps and carefully pass a small hook through it. Try not to rip a gaping hole in the ventricle—take your time and be careful. Keep the heart moist at all times with frog Ringer solution. However, do not moisten the heart during periods of data recording. Note that the heart is beating all by itself, thus dispensing with the need for a stimulator in most parts of this experiment.

Kymograph Procedure

Set the speed control knob on the kymograph to position 3. Arrange the heart lever and signal magnet on the kymograph support so that each pen tip is writing on the same vertical line of the drum. Attach a piece of fine thread to the hook that has been passed through the ventricle and tie it to the end of the heart lever. The heart lever should be horizontal to the lab bench and the thread should be vertical. This will insure that each heartbeat will give a maximal tracing on the drum. Turn on the recorder.

Physiograph Procedure

Set the paper speed to 1 cm/sec. Position the pen on the paper near the bottom of its channel so that large pen deflections will remain on the paper. Attach a piece of fine thread to the hook that has been passed through the ventricle and tie it to the hook on the myograph. The thread should be vertical. Adjust the myograph tension adjuster until the thread is taut. Start the paper moving.

8.1. *Normal Heartbeat.* Obtain a record of six to eight heartbeats. Compare your recording with Figure 8.2. Note that each beat consists of two bumps—a small atrial beat and a large ventricular beat. The frog heart has three chambers—two atria which give rise to the small bump and one ventricle which gives rise to the large bump.

88

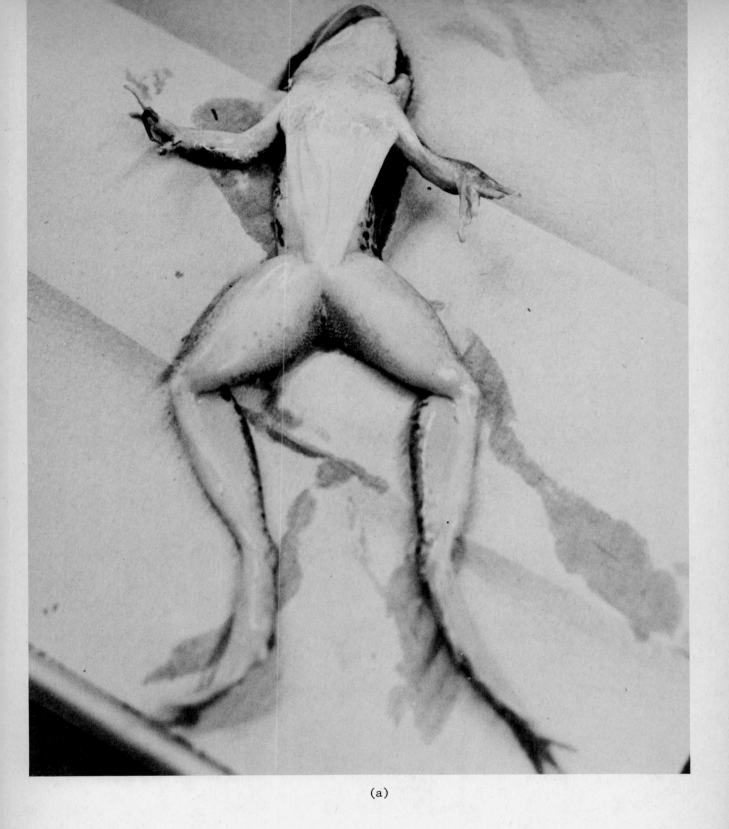

Figure 8.1 Dissection of the frog heart. (a) Frog with ventral surface facing
upward.

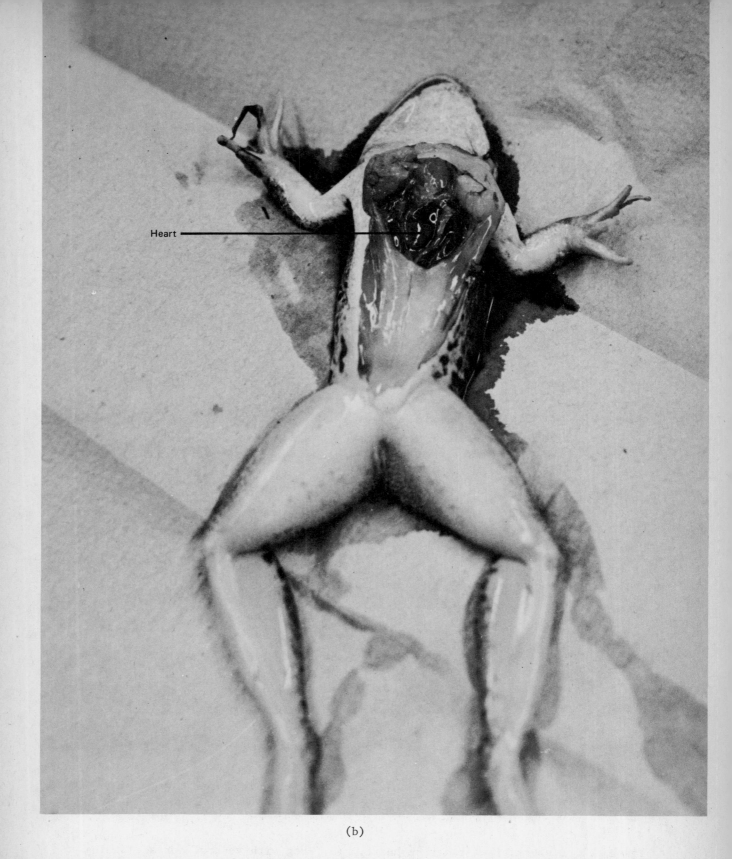

Heart

(b)

Figure 8.1 (Continued) Dissection of the frog heart. (b) Location of the heart.

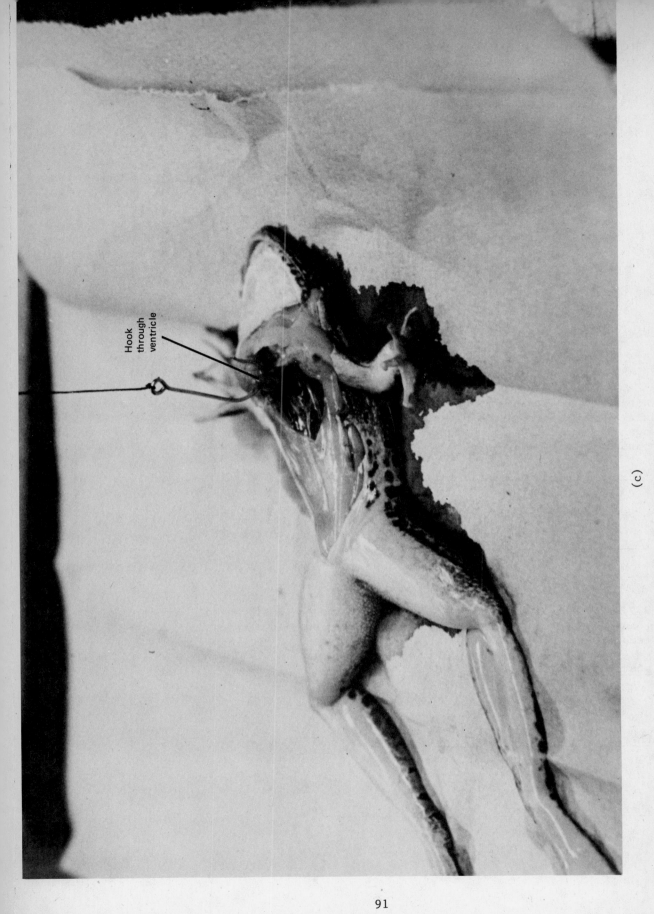

Hook
through
ventricle

Figure 8.1 (Continued) Dissection of the frog heart. (c) Hook placed in ventricle.

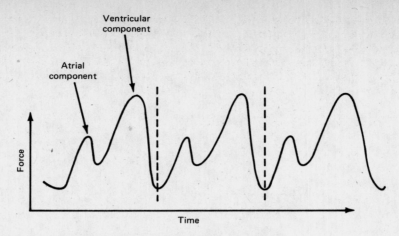

Figure 8.2 A typical frog heartbeat pattern.

1. Why is the ventricular part of the record larger?

    Note that the heart is contracting during the upward phase of the record and relaxing during the downward phase of the record. Allow the heart to beat for 15 seconds and then count the number of ventricular beats that have occurred. Multiply this number by four to obtain the resting heart rate in beats per minute.

2. What is the resting heart rate?

3. What is the resting ventricular amplitude?

    8.2. Effect of Temperature. Slowly apply some cold Ringer solution on the heart and observe the change in heart rate and contraction amplitude. Measure the temperature of the Ringer with a thermometer.

4. What is the temperature of the cold Ringer?

5. What is the heart rate with cold Ringer?

6. What is the ventricular amplitude with cold Ringer?

Apply normal room-temperature Ringer (25°C) to the heart until control heart rate and amplitude have been obtained. Then apply some warm Ringer to the heart. Measure the temperature of the warm Ringer before applying it to the heart. Ringer warmer than 40°C will denature proteins in the heart and will cause irreversible damage to the preparation. After applying the warm Ringer, observe any changes in heart rate and contraction amplitude.

7. What is the temperature of the warm Ringer?

8. What is the heart rate with warm Ringer?

9. What is the ventricular amplitude with warm Ringer?

Then apply normal room-temperature Ringer to the heart until control heart rate and contraction amplitude have been obtained. The temperatures that you have just experimentally imposed are in fact encountered by frogs living in a pond. It is interesting to note that frogs and other cold-blooded animals really are at the mercy of their environment.

8.3. Effect of Hormones. It is possible to mimic the action of autonomic nerves on the frog heart by applying the neurotransmitters released by these nerves. Acetylcholine is released by parasympathetic fibers in the vagus nerve; epinephrine is released by sympathetic fibers in the cervical ganglia.

Apply 2 drops of 1-mM ACh dissolved in Ringer to the heart. Watch the heart and when you see a change in the beating, run a short record. If nothing has happened in 2 minutes, add 2 additional drops of ACh *but no more*. If the heart stops beating, return it to control levels by rinsing it with warm Ringer.

10. What is the heart rate after ACh application?

11. What is the ventricular amplitude after ACh?

Wait 1 minute and then run another record.

12. Has the heart begun to recover?

Rinse the heart thoroughly with control Ringer and wait for 5 minutes before proceeding. Now rinse the heart with two drops of 100-mM epinephrine dissolved in Ringer. Run a short record.

93

13. What is the heart rate after epinephrine application?

14. What is the ventricular amplitude after epinephrine?

Rinse the heart thoroughly with control Ringer until control values are obtained. It may take longer to return to control levels because the effects of epinephrine wear off more slowly than the effects of ACh. Then add 5 drops of 1-mM atropine dissolved in Ringer. Wait 30 seconds and then run a short record.

15. What is the heart rate after atropine application?

16. What is the ventricular amplitude after atropine?

Now add 2 to 4 drops of 1-mM ACh dissolved in Ringer and run a record.

17. What is the heart rate after ACh?

18. What is the ventricular amplitude after ACh?

Rinse the heart with control Ringer again. The frog heartbeat originates in a small area called the SV node. In this area, the muscle membrane shows a slow inward Na^+ leakage. This leads to a gradual depolarization of the membrane and spontaneous action potential generation. ACh exerts its effects by increasing the outward K^+ leak so as to oppose the Na^+ leak and thus hyperpolarize the membrane. As a result, the time required for the Na^+ leak to depolarize the membrane to threshold is lengthened.

Epinephrine exerts its effects both at the SV node and over the entire heart. It increases the Na^+ permeability and affects the contractile process itself by increasing the force of contraction.

Atropine combines with ACh receptors and competitively inhibits the ACh-receptor interaction.

8.4. Origin of the Heatbeat. To demonstrate that the ventricular beat originates in the atria, tie a piece of fine thread around the heart between the atria and the ventricle, as shown in Figure 8.3. In classical physiology, this is called a "second Stannius ligature," after Stannius, the physiologist who first tied it. This ligature pinches the specialized conducting cells that transport the excitation long enough for the atria to complete their contraction before the ventricle is excited. These cells, with the ligature in place, may not conduct at all or they may conduct so slowly that they transmit only one action potential for every two to

94

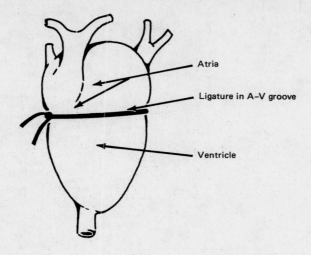

Figure 8.3 Placement of ligature for heart block.

three atrial contractions. This is called *heart block*. After you have caused heart block, the ventricular bump of the record should have disappeared. It should still be possible to stimulate the ventricle to beat by means of an electrical stimulus. Use a single stimulus pulse of 10-V amplitude and 1-msec duration to stimulate the heart.

<u>19</u>. <u>Does the ventricle contract after electrical stimulation</u>?

If you have been careful and lucky, the thread tied around the heart has only blocked the coronary circulation and has stopped conduction by anoxia rather than by mechanical damage to the cells. If this is the case, you may be able to restore normal conduction by removing the thread.

<u>20</u>. <u>Can you restore normal conduction through the heart</u>?

If your heart has recovered, remove it from the frog and place it in a beaker of Ringer solution.

<u>21</u>. <u>Does the heart continue to beat after removal from the frog</u>?

Name _____ Lab Section _____ Date _____

8.1. Normal Heartbeat

1. Why is the ventricular part of the record larger?

   _____

   _____

   _____

2. What is the resting heart rate? _____

3. What is the resting ventricular amplitude? _____

   Sample Record—Normal Heartbeat

8.2. Effect of Temperature

4. What is the temperature of the cold Ringer? _____

5. What is the heart rate with cold Ringer? _____

6. What is the ventricular amplitude with cold Ringer? _____

    Sample Record—Cold Ringer

7. What is the temperature of the warm Ringer? _____

8. What is the heart rate with warm Ringer? _____

9. What is the ventricular amplitude with warm Ringer? _____

    Sample Record—Warm Ringer

8.3. Effect of Hormones

10. What is the heart rate after ACh application? _____

11. What is the ventricular amplitude after ACh? _____

98

12. Has the heart begun to recover? _____

    Sample Record—ACh

13. What is the heart rate after epinephrine application? _____

14. What is the ventricular amplitude after epinephrine? _____

    Sample Record—Epinephrine

15. What is the heart rate after atropine application? _____

16. What is the ventricular amplitude after atropine? _____

Sample Record—Atropine

17. What is the heart rate after ACh? _____

18. What is the ventricular amplitude after ACh? _____

Sample Record—Atopine + ACh

8.4. Origin of the Heartbeat

19. Does the ventricle contract after electrical stimulation? _____

20. Can you restore normal conduction through the heart? _____

21. Does the heart continue to beat after removal from the frog? _____

100

Sample Record—Heart Block

EXPERIMENT 9

HEMATOLOGY

Behavioral Objectives

The student should be able to:

1. State the functions of blood.
2. Perform a hemoglobin determination using the cyanmethemoglobin method.
3. Perform a hematocrit determination on a blood sample.
4. Determine the blood type of a blood sample.
5. Determine the bleeding and clotting time of a blood sample.

Materials

Sterile blood lancets (or Auto-Let)
Sterile cotton
Alcohol swabs
Microscope slides
Wax marking pencils
Microhematocrit centrifuge
Microhematocrit reader
Rh typing boxes
Wood applicator sticks (plain)
Heparinized and nonheparinized capillary tubes
20 µl micro-sampling pipettes
Beakers (50 ml)
Files (triangular)
Spectronic 21 spectrophotometer
Spectronic 21 cuvettes
Seal-ease
Pipettes (5 ml) with dispensing bulbs
Drabkin's reagent
Cyanmethemoglobin standard solution
Anti-A blood serum
Anti-B blood serum
Anti-D blood serum

Introduction

Blood is a type of connective tissue that contains dispersed cells called *formed elements* dispersed in a liquid matrix called *plasma*. Blood makes up about 8 percent of body weight and has a normal volume of about five liters. It has an average pH of 7.4 and is 4.5 to 5.5 times thicker than water. The blood transports materials to and from cells of the body tissues, helps to regulate body temperature and pH, protects the body from infection, and prevents the excessive loss of body fluids.

Blood plasma is 90 percent water and contains numerous substances dissolved in it: proteins (albumin, globulins, and so on), waste products (urea, uric acid), nutrients (glucose, amino acids, fatty acids), regulatory substances (hormones, enzymes), gases (oxygen, carbon dioxide), and various ions (Na^+, K^+, Ca^{++}, Mg^{++}, Cl^-, PO_4^{-3}, SO_4^{-2}, HCO_3^-). The *erythrocytes* (red blood cells) are the most numerous formed elements and contribute to the blood viscosity. Erythrocytes contain large amounts of *hemoglobin*, a respiratory pigment, which increases the oxygen-carrying capacity of the blood about 70 times. Hemoglobin is a conjugated protein molecule which contains four heme groups; each heme group can reversibly bind to a molecule of oxygen. Hemoglobin becomes saturated with oxygen when erythrocytes are traveling through capillaries in the lungs and releases oxygen to tissue cells when they are traveling through capillaries in the tissues of the body. *Leucocytes* (white blood cells) are much less numerous than erythrocytes (one white cell for every 700 red cells) and lack hemoglobin. There are five common types of white blood cells: neutrophils, eosinophils, basophils, lymphocytes, and monocytes. The general function of these leucocytes is to protect the body from infection. *Thrombocytes* (platelets) are the least numerous and smallest of the formed elements. They are cell fragments which are surrounded by a membrane. Thrombocytes function in initiating the blood clotting process.

Experimental Procedure

In this experiment, blood will be obtained by the following finger-puncture procedure:

1. Use a finger other than your thumb or index finger for the puncture.
2. Clean the finger of choice with an alcohol swab. This will not only clean the finger thoroughly but will also produce an increase in blood flow to the fingertip.
3. Let the finger air dry and, using a blood lancet or an Auto-Let, make a puncture in the tip of the cleansed finger.
4. Discard the first drop of blood by wiping it off with a ball of sterile cotton. This drop is discarded because it is contaminated with tissue fluid. Use the second and subsequent drops for the procedures outlined below. Avoid squeezing the finger as this also contaminates the blood with tissue fluid.
5. If the puncture continues to bleed after enough blood has been obtained, hold a ball of cotton over the puncture and apply pressure until the bleeding stops.
6. At the end of the lab period, dispose of all lancets, microscope slides, and so on, that have been in contact with the blood.

9.1. Hemoglobin Determination.

A variety of methods are used to determine the amount of hemoglobin in blood. These methods can be categorized as colorimetric, gasometric, chemical, or gravitational. The *cyanmethemoglobin method*, a chemical-colorimetric method, is the most accurate and most often used. The test involves reacting hemoglobin with potassium cyanide to form a colored compound, cyanmethemoglobin. The concentration of the cyanmethemoglobin, and thus the concentration of hemoglobin, is then determined spectrophotometrically.

Fill a 20-μl micro-sampling pipette with 20 μl of blood. Fill the tube to the black mark and then empty the tube into a beaker containing 5 ml of Drabkin's reagent. Mix the contents of the beaker and allow it to stand for 2 to 3 minutes.

Note that Drabkin's reagent contains KCN which is deadly poison and extremely toxic. Be careful not to pipette it by mouth or spill it!

While the reaction is occurring, it will be necessary to standardize the Spectronic 21 spectrophotometer. Adjust the knob on the machine until it is set to absorb light at 540 nm. Fill a cuvette with an aliquot of Drabkin's reagent. Place this cuvette in the sample holder so that the white mark on the cuvette lines up with the mark on the machine. Using the other knob on the machine, set the spectrophotometer to read 100 percent transmittance.

Using the standard solutions provided by your instructor, determine the percent transmittance for each solution of a known hemoglobin concentration. Use the special piece of semilog paper on the data sheet to plot percent transmittance versus hemoglobin concentration to obtain a standard curve for your machine.

Then place an aliquot of the unknown sample in a cuvette and read its percent transmittance. Using the standard curve, find the hemoglobin concentration that this corresponds to on the standard curve.

1. What is the hemoglobin percentage of your sample in grams percent?

Average values are 15 to 16 g% for males and 14 g% for females.

9.2. Hematocrit Determination. The blood volume of an average person is between 5 and 6 liters. The percentage of this volume that is occupied by red blood cells is called the *packed cell volume* (PCV) or *hematocrit*. The hematocrit accounts for approximately 45 percent of the total blood volume. The determination of hematocrit, along with a hemoglobin determination, provides a rapid screening method for anemias.

Obtain a drop of blood and place it on a microscope slide. Insert one end of a *heparinized* capillary tube into the drop. Hold the tube horizontally and slightly downward to allow the blood to fill the tube about two-thirds full. Avoid getting air bubbles in the tube if possible. Seal the clean end of the tube with seal-ease. Place the sealed tube in one of the radial grooves of the microhematocrit centrifuge, with the sealed end against the outer rim. Load the centrifuge with an even number of tubes and place them in opposite slots so that the centrifuge head is properly balanced. Secure the inside cover and fasten down the outside cover.

Centrifuge the sample for 5 minutes. Then remove your capillary tube and identify the various components of the blood by comparing the tube with Figure 9.1.

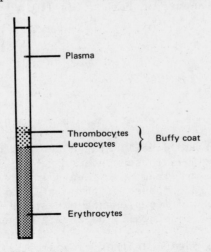

Figure 9.1 Typical microhematocrit tube after centrifugation.

Determine the hematocrit by placing the tube in the hematocrit reader. Instructions for reading the hematocrit are given on the reader. Normal values are 40 to 54 percent (average = 47 percent) in men and 37 to 47 percent (average = 42 percent) in women.

2. What is your hematocrit?

9.3. Blood Type Determination. There are four blood types possible in man: A, B, O, and AB. The blood types are named according to the antigens present on the surface of the red blood cell membrane. A and B antigens are composed of an amino acid complex in combination with a polysaccharide. The presence or absence of these antigens depends on the expression of an inherited pair of genes. Although there are six possible genotypes, as shown in Table 9.1, only four phenotypes can be distinguished because O is recessive. Whichever antigen is found on the red blood cell, its reciprocal antibody is found in the plasma.

Obtain a microscope slide and draw a line down the middle of it with a wax marking pencil. Place the slide on top of a paper towel and label the left side of the slide "anti-A" and the right side of the slide "anti-B." Then place a drop of anti-A serum on the left side of the slide and a drop of anti-B serum on the right side of the slide. Add a drop of blood to each of these drops. Mix the combined drops with a wood applicator stick and watch for *agglutination* (clumping).

Table 9.1 Genotypes of Blood Groups

TYPE	ANTIGEN	ANTIBODY	GENOTYPE	% POPULATION
A	A	Anti-B	AA, AO	42
B	B	Anti-A	BB, BO	10
AB	AB	None	AB	3
O	None	Anti-A, Anti-B	OO	45

Agglutination is caused by the formation of insoluble antigen-antibody complexes. If agglutination occurs on the left side, the blood is type A; if it occurs on the right side, it is type B; if it occurs on both sides it is type AB; if it doesn't occur at all, it is type O.

3. What is your blood type?

The frequencies of occurrence of the various blood types in selected populations are summarized in Table 9.2.

The *Rh factor* is another red blood cell antigen that is found in about 85 percent of the population. The "Rh" name was chosen because rabbit blood was being transfused into *Rhesus* monkeys in the original experiments involving the Rh factor. The Rh factor is also inherited and is composed of three allelic genes (*Cc, Dd, and Ee*). The *D* gene is the most important and its presence makes red blood cells Rh^+. Neither Rh^+ nor Rh^- blood possesses naturally occurring Rh antibodies in the plasma.

To determine if you are Rh^+, obtain another microscope slide. Place a drop of anti-D serum on the slide and add a drop of blood. Mix the combined drop with a

Table 9.2 Blood Groups in Selected Populations

| | FREQUENCY (%) | | | |
POPULATION	O	A	B	AB
U.S. whites	45	41	10	4
U.S. blacks	47	28	20	5
Australian aborigines	31	66	0	0
Pure Peruvian Indians	100	0	0	0

wood applicator stick and place the slide on an Rh-typing box. Turn on the light and gently tilt the box back and forth. If agglutination occurs, the blood is Rh$^+$.

4. What is your Rh blood type?

9.4. Bleeding Time Determination. The platelets and some components of the plasma are needed for blood clotting. Platelets prevent blood loss by initiating a series of reactions that result in a blood clot. The *bleeding time* is the time required for a small incision to stop bleeding. It is a measure of adequate platelet number and function.

Puncture your ear lobe or finger tip and record the time. At 30-second intervals, lightly touch the blood coming from the puncture with a ball of sterile cotton. Repeat this procedure until the blood stops flowing. Normal values are between 1 and 3 minutes.

5. What is your bleeding time?

9.5. Clotting Time Determination. Obtain a *nonheparinized* capillary tube and fill it two-thirds full with blood. Fill the tube as was done for the hematocrit determination. After 1 minute, use a triangular file to score the tube. Carefully break off a small (about 1 cm) piece of the tube and determine if a thread of clotted blood can be seen between the broken pieces. Continue scoring and breaking the tube every 30 seconds until the threads are visible.

6. What is the clotting time of your blood sample?

DATA SHEET FOR EXPERIMENT 9: HEMATOLOGY

Name _____ Lab Section _____ Date _____

9.1. Hemoglobin Determination

Standard	% Transmittance (540 nm)
1	
2	
3	
Unknown	

1. What is the hemoglobin percentage of your sample in grams percent?

_____

9.2. Hematocrit Determination

2. What is your hematocrit? _____

9.3. Blood Type Determination

3. What is your blood type? _____

4. What is your Rh blood type? _____

9.4. Bleeding Time Determination

5. What is your bleeding time? _____

9.5. Clotting Time Determination

6. What is the clotting time of your blood sample?

_____

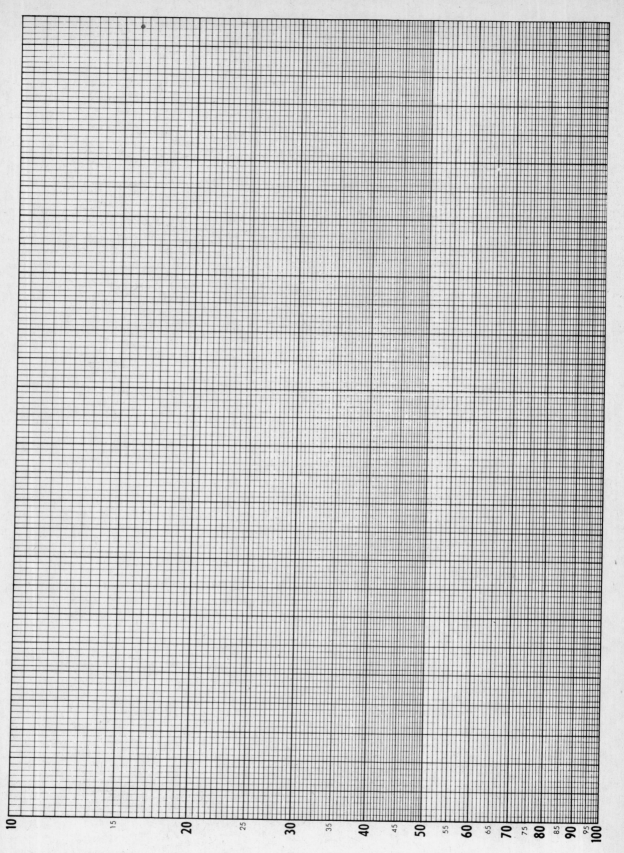

10
15
20
25
30
35
40
45
50
55
60
65
70
75
80
85
90
95
100

110

EXPERIMENT 10

HUMAN ELECTROCARDIOGRAM

Behavioral Objectives

The student should be able to:

1. List the parts of the cardiac conduction system.
2. State the arrangement of the limb and chest leads used to measure an ECG.
3. Identify the important parts of an ECG tracing.
4. Record a phonocardiogram.
5. Correlate the time relationships between the ECG and the phonocardiogram.
6. Define tachycardia, bradycardia, and arrhythmia.
7. Distinguish between clean and noisy ECG records.
8. Calculate the electrical axis of the heart from ECG data.
9. List the disorders associated with right and left axis deviation.

Materials

Plastic rulers
Physiograph with ECG coupler
ECG patient cable
ECG limb and chest electrodes
Heart sounds microphone
ECG electrode paste
Alcohol swabs

Introduction

A recording of the electrical activity of the heart is called an *electrocardiogram* (ECG, EKG). The ECG is a series of electrical waves produced by the excitation of the heart. In the human heart, this excitation originates in an area of the right atrium called the *sinoatrial (S-A) node*. Thus the S-A node acts as a pacemaker and determines the heart rate and rhythm. The excitation initiated by the S-A node spreads down the atria and activates the *atrioventricular (A-V) node*, another specialized area of the heart near the junction of the right atrium and right ventricle. The excitation wave is then carried into the ventricles by the *bundle of His*. Once inside the ventricles, the bundle of His branches to form a network of *Purkinje fibers* that conduct the excitation to all parts of the ventricles.

Since the body is composed of conducting fluids, electrodes placed on the surface of the body are capable of recording the electrical activity of the heart.

The ECG can be recorded by attaching electrodes to the right and left arms and the left leg. An exploring (moveable) electrode may also be attached to six different locations on the chest. Three sets of electrode ("lead") positions have been adopted in an effort to standardize the recordings made by electrocardiographers around the world.

   10.1.   Bipolar (Limb) Leads. The bipolar limb leads are attached to the wrists and ankles. These leads are called "bipolar" because they record the difference in voltage between *two* points on the body:

Lead I—left arm and right arm
Lead II—left leg and right arm
Lead III—left leg and left arm

   The principles underlying the use of these electrode positions were developed by Willem Einthoven, a Dutch physiologist, who is recognized as the "Father of Electrocardiography." Einthoven postulated that the electrical field of the heart could be represented on the frontal plane of the body as an electrically equilateral triangle, with the heart at its center and the extremities as its apices (Figure 10.1).

   A wave of depolarization advancing toward a positive electrode produces an upward (+) pen deflection on the electrocardiogram tracing. The magnitude of the deflection is directly proportional to how closely the lead axis parallels the direction of the electrical force. In Figure 10.2, Lead II would be expected to show a larger pen deflection than Lead III, even though both are recording the same heartbeat. This is because Lead II is almost parallel to the direction of electrical force, whereas Lead III is almost perpendicular to the direction of electrical force.

   When the sides of the Einthoven triangle are translated to the location of the heart (the center of the triangle), a triaxial reference system is produced (Figure 10.3). Polarity of the leads is indicated by + and − signs. In other words, in Lead I, voltage changes occur in both left and right arm fields, and the difference between these voltages is recorded.

   10.2.   Unipolar (Augmented Limb) Leads. To increase the size of a given pen deflection, it is desirable to record these voltage differences with respect to ground, rather than with respect to each other. Since there is no single point on the body surface that is unaffected by the electrical activity of the heart, a close approximation to ground is used. This is done by connecting the leads from the left arm, right arm, and left leg together to form a "floating ground" or "central reference terminal." Then, by electronically separating the lead of

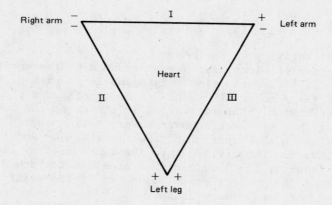

Figure 10.1   The Einthoven triangle.

112

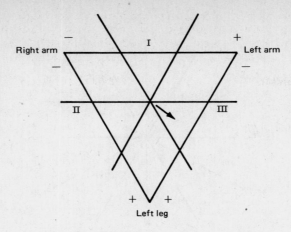

Figure 10.2

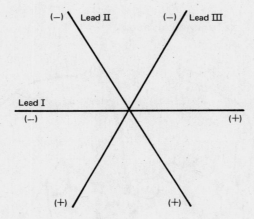

Figure 10.3   The triaxial reference system.

interest from the central reference terminal, the difference in voltage between one point on the body and ground can be recorded:

Lead aVr—right arm and ground
Lead aVl—left arm and ground
Lead aVf—left leg and ground

These three leads are called "unipolar" for the reasons mentioned above. They are also referred to as "augmented" limb leads because they show larger pen deflections than the bipolar leads. When these unipolar leads are plotted on the Einthoven triangle and are translated to its center, a hexaxial reference system results, as shown in Figure 10.4.

  10.3. Precordial (Chest) Leads. A third set of leads has been developed because of the need to detect changes in electrical activity of the heart which are not confined to the frontal plane of the body. These leads measure the voltage difference between the central reference terminal and an exploring electrode placed on one of six chest locations (Figure 10.5). These leads are also unipolar leads. They are sometimes called "precordial" leads because of their location—the precordia—upper central (epigastric) region of the abdomen and anterior surface of the thorax.

113

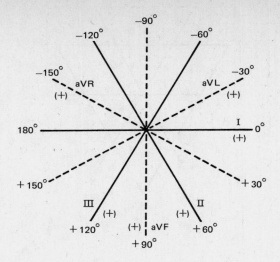

Figure 10.4   The hexaxial reference system.

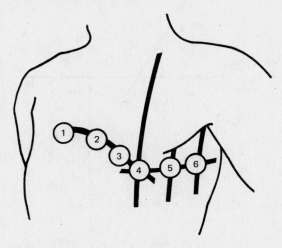

Figure 10.5   Placement of precordial (chest) ECG electrodes.

   10.4.   The ECG Tracing.   The various waves of an ECG tracing are designated by
the letters P, Q, R, S, T, and U.   ECG paper is ruled in lines which are 1 mm apart
both horizontally and vertically.   Horizontally, the lines indicate change in
time—each 1-mm space representing 0.04 second.   Vertically, the lines indicate
change in voltage—each 1-mm space representing 0.1 mV.   Heart rate may be calculated
by counting the number of R to R intervals in 3 seconds and then multiplying this
number by 20 to get beats per minute.
   A typical ECG tracing recorded from Lead II is shown in Figure 10.6.   Typical
amplitudes and times for each of the ECG waves are summarized in Table 10.1.   An
ECG shows the following characteristic parts:

P wave—produced by atrial depolarization.
P-R interval—the time between the start of the P wave and the start of the R wave;
   represents the time for the impulse to move from the S-A node across the atria,
   through the A-V node, bundle of His, and Purkinje fibers.
QRS complex—produced by ventricular depolarization.

114

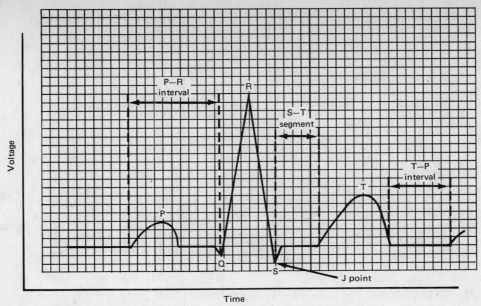

Figure 10.6 Typical ECG tracing.

Table 10.1 Typical ECG Amplitudes and Durations

PHASE	AMPLITUDE (mV)	DURATION (sec)
P wave	0.2	0.08
P-R interval	--	0.16
QRS complex	0.8-1.2	0.04-0.09
S-T segment	--	0.12
T wave	0.3	0.16

S-T segment—the time between the end of the R wave and the beginning of the T wave; represents the time between completion of ventricular depolarization and beginning of repolarization.

J point—the junction between the QRS complex and the S-T segment.

T wave—produced by repolarization of the ventricles.

T-P interval—the interval between the end of the T wave and the start of the next P wave; the time during which the heart is recovering between cardiac cycles.

U wave—follows the T wave; usually small or not detectable; caused by an after-potential, possibly of a papillary muscle.

10.5. The Phonocardiogram. There are at least two distinct heart sounds that accompany each heartbeat. The first sound ("lubb") is of low pitch and long duration. It is produced by closure of the A-V valves at the beginning of ventricular systole, the opening of the semilunar valves, the flow of blood into the pulmonary trunk and aorta, and the contraction of ventricular muscle. The second sound ("dup") is of higher pitch and shorter duration. It is produced by the closing of the semilunar valves and occurs at the beginning of ventricular diastole (Figure 10.7). In a slowly beating heart, a third heart sound may be heard. It is dull, low-pitched, and is thought to be caused by vibration of the A-V valves when blood rushes into the incompletely filled ventricles. The third heart sound is considered normal in children and adults under age 35. When heard

115

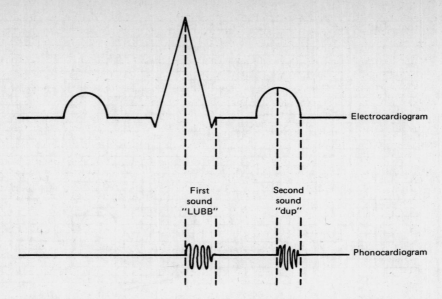

Figure 10.7 Relationship of the electrocardiogram and phonocardiogram.

in older people, it is often a sign of cardiac failure. A fourth sound, rare in normal people, may be heard in cases of hypertensive cardiovascular disease, aortic stenosis, or the like. It is due to atrial contraction generated by forceful fluttering of the A-V valves.

A recording of the heart sounds is called a *phonocardiogram*. Phonocardiograms are used to diagnose heart murmurs. If a heart valve is supposed to be closed but remains open for some reason, a regurgitation of blood will occur. This leads to pumping inefficiency and an increase in cardiac work. If a valve is supposed to be open but remains partially closed (stenosed), greater pressures are needed to force blood through this valve. Murmurs are classified as systolic ("lubb-sschtup") or diastolic ("sschlubb-dup"), depending on when they occur:

Systolic murmurs
    Aortic stenosis—semilunar valves are supposed to be open but are partially closed.
    Mitral regurgitation—A-V valves are supposed to be closed but are partially open.
Diastolic murmurs
    Aortic regurgitation—semilunar valves are supposed to be closed but are partially open.
    Mitral stenosis—A-V valves are supposed to be open but are partially closed.

10.6. Cardiac Arrhythmias. The normal adult human heart beats at a rate of 60 to 100 beats per minute at rest. Heart rates greater than 100 beats per minute are called *tachycardia*; heart rates less than 60 beats per minute are called *bradycardia*. Under normal conditions, the time interval between beats remains constant and the rhythm is termed "regular." If there is a variation in the time between successive beats, the rhythm is termed "irregular." Arrhythmias occur as a result of changes in the pacemaker, conduction disturbances, or both.

Experimental Procedure

Electrocardiograms that are taken at rest are always done with the person lying down. This is because different lead deflections will occur if the position of

116

the heart inside the body changes. Have your partner lie down on the lab bench and try to relax. Rub the inside surface of your partner's calves and forearms with an alcohol swab. This will remove some of the keratinized skin found in these areas and will lower skin resistance. Apply a small amount of ECG paste to your partner's left arm, right arm, left leg, and right leg. Attach a plate electrode to each limb by means of a rubber strap. Make sure that the straps are snug but not so tight that circulation of blood is reduced. Attach the ECG wires to the plate electrodes by using the thumb screws. These wires are stamped with an abbreviation and are color-coded as shown in Table 10.2.

To record the phonocardiogram, attach a heart-sounds microphone to your partner in the vicinity of the heart. Connect the microphone wire to the input of the DC–AC or high-gain coupler of the physiograph. Since the physiograph is equipped with two recording channels, the ECG and the phonocardiogram will be displayed simultaneously on the chart paper. This display will permit direct comparison of the relative time relationships between the two events.

Leads I, II, III, aVr, aVl, aVf, and the chest (V) leads will be used. Turn the lead selector switch on the cardiac coupler to the Lead I position. Run the paper speed at 2.5 cm/sec for all recordings. Set the time marker to provide one mark per second. Set the sensitivity of the phonocardiogram channel amplifier to 100 mV/cm and set the filter to 100 Hz. Set the gain of the DC–AC coupler to 10 X and set the time constant to 0.3. Start the physiograph and obtain a short record. Compare this record with the records shown in Figure 10.8. If the electrodes and microphone have been applied correctly, and your partner and the physiograph are cooperative, there is no reason to obtain a noisy record. Once a "clean" ECG and phonocardiogram have been obtained, switch the cardiac coupler to Lead II and return the pen to baseline with the reset switch. Run a short record of Lead II, and then repeat the procedure for Leads III, aVr, aVl, and aVf. Then attach a chest electrode to the V wire of the ECG patient cable. Place small blobs of ECG paste in the locations shown in Figure 10.5. Run records for each of the six chest leads by setting the cardiac coupler to the V setting and then moving the chest electrode to the six different chest locations. The records that you obtain should be similar to those shown in Figure 10.9.

1. <u>Choose a lead in which you have a clear phonocardiogram. With respect to the ECG, when do the first and second heart sounds occur?</u>

2. <u>What is the resting heart rate of your partner?</u>

Table 10.2 ECG Lead Designations

WIRE	ABBREVIATION	COLOR
Right arm	RA	White
Left arm	LA	Black
Right leg	RL	Green
Left leg	LL	Red

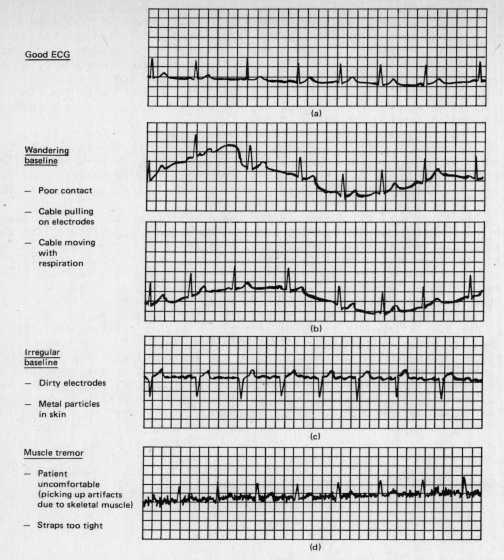

Good ECG

(a)

Wandering
baseline

— Poor contact

— Cable pulling
on electrodes

— Cable moving
with
respiration

(b)

Irregular
baseline

— Dirty electrodes

— Metal particles
in skin

(c)

Muscle tremor

— Patient
uncomfortable
(picking up artifacts
due to skeletal muscle)

— Straps too tight

(d)

Figure 10.8 Artifacts associated with the ECG. (From H. E. Thomas, *Handbook of Biomedical Instrumentation and Measurement*, 1974, p. 44. Reprinted with permission of Reston Publishing Co., a Prentice-Hall Co., 11480 Sunset Hills Road, Reston, VA 22090.)

10.7. Analysis. The electrical activity of the heart produces several potentials that differ in force and direction in a three-dimensional field. The electrical force of the heart at any given moment is the sum of all of the positive and negative forces present at that moment. Such a resultant force, having a direction as well as a magnitude, is a vector quantity. A vector quantity may be represented by an arrow whose length is proportional to the magnitude of the force and whose direction indicates the direction of the force.

It is possible to construct vectors for each ECG wave, but it is usually done only for the QRS complex. This vector is given a special name—the *electrical axis of the heart*—since it represents the sum of the forces acting on the ventricles. In a normal heart, the electrical axis lies between 0 and +90 degrees. An electrical axis between 0 and -90 degrees indicates left-axis deviation; an electrical axis

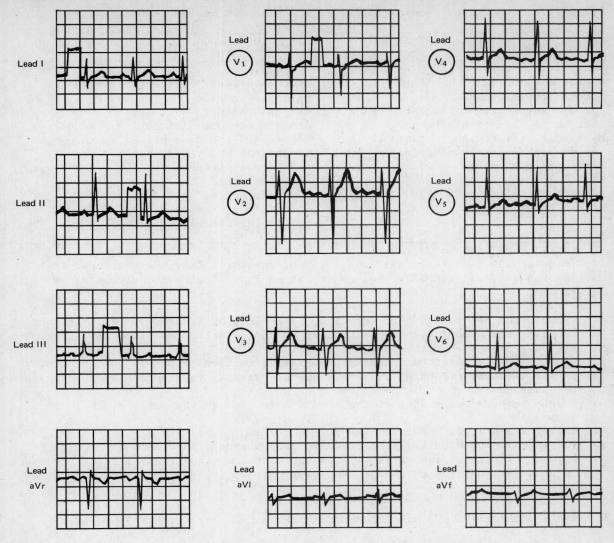

Figure 10.9   Typical ECG recordings.   (From H. E. Thomas, *Handbook of Biomedical Instrumentation and Measurement*, 1974, p. 51.   Reprinted with permission of Reston Publishing Co., a Prentice-Hall Co., 11480 Sunset Hills Road, Reston, VA 22090.)

between +90 and +180 degrees indicates right-axis deviation.   Left- or right-axis deviation *may* indicate a structural abnormality (for example, ventricular hypertrophy) or a pathogenic state (Table 10.3).

The normal mean QRS vector (electrical axis) varies greatly from person to person.   Age, body build, and position of the heart can all influence the QRS axis of a normal heart.   From birth to 1 year of age, the QRS axis is between +90 and +150 degrees.   From ages 1 to 8, the axis is usually between +45 and +105 degrees. From age 8 to adulthood it is usually between 0 and +90 degrees.   This normal axis variation is explained by the fact that the right ventricle is large relative to the left ventricle in children.   With increasing age, the predominance of the left ventricle deviates the QRS axis to the left.

Table 10.3 Conditions Leading to Electrical Axis Deviations

LEFT-AXIS DEVIATION	RIGHT-AXIS DEVIATION
Elevated diaphragm	Normal vertical heart
Chronic hypertension	Mitral stenosis
Chronic aortic regurgitation	Cor pulmonale
Left bundle branch block	Acute myocardial infarction
Myocardial infarction	Right bundle branch block
Interventricular septal defect	Congenital heart defects:
Coarctation of aorta	Pulmonary stenosis
Mitral insufficiency	Dextrocardia
	Ostium primum septal defect
	Ebstein's disease
	Tetralogy of Fallot

To determine the electrical axis for your partner's ECG, use the recordings made in Leads I, II, and III. Follow the steps listed below to calculate the mean QRS vector:

1. Measure the *net* positive or negative QRS deflection in Lead I. This is done by counting the number of paper divisions up and down from baseline and then subtracting them from each other.
2. Plot the net deflection in the appropriate direction (+ or −) from the center of the axis corresponding to Lead I on the triaxial reference system shown on the data sheet.
3. Draw a perpendicular line to the Lead I axis through the point plotted in step 2 above.
4. Repeat steps 1, 2, and 3 for Leads II and III.
5. The three perpendicular lines drawn in step 3 should intersect at a point to form a triangle. Draw an arrow from the center of the triaxial reference system to this intersection or to the center of this triangle. This is the mean QRS axis.

<u>3.</u> <u>Is the electrical axis in the normal range? Why?</u>

To determine the amplitude and duration of the waves of your partner's ECG, use the physiograph settings and Lead II recording to complete the table given on the data sheet.

Calculate the mean QRS axis for the patient described on the data sheet.

<u>4.</u> <u>To which patient do you tentatively assign the records? Why?</u>

Name _____ Lab Section _____ Date _____

Normal ECG

Sample Record—Lead I

Sample Record—Lead II

Phase	Amplitude (mV)	Duration (msec)
P wave		
P-R interval	------	
QRS complex		
T wave		

Sample Record—Lead III

Sample Record—Lead aVr

Sample Record—Lead aVl

Sample Record—Lead aVf

Sample Record—Lead V1

Sample Record—Lead V2

Sample Record—Lead V3

Sample Record—Lead V4

Sample Record—Lead V5

124

Sample Record—Lead V6

Sample Record—Phonocardiogram (Lead I)

1. With respect to the ECG, when do the first and second heart sounds occur?

    S1_____

    S2_____

2. What is the resting heart rate of your partner?

    _____

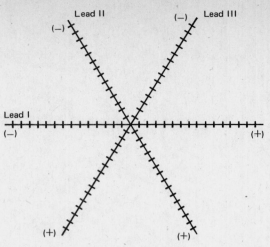

3. Is the electrical axis in the normal range? _____

    Why? _____

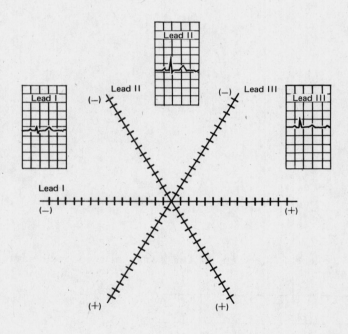

Lead I, II, and III recordings from an unknown person are shown above. Plot the mean QRS vector (electrical axis) on the triaxial figure provided.

The above records may belong to:

a. A male clerk, age 60, height 5 feet 9 inches, 196 pounds, in for an insurance examination.

b. A male computer programmer, age 26, height 6 feet 2 inches, 155 pounds, complaining of eye strain.

c. A female orthodontist, age 30, height 5 feet 3 inches, 120 pounds, having a routine checkup in her eighth month of pregnancy.

4. To which person do you tentatively assign the records?

_____

Why? _____

_____

_____

_____

EXERCISE PHYSIOLOGY

Behavioral Objectives

The student should be able to:

1. Distinguish between systolic and diastolic blood pressure.
2. Identify the normal range of blood pressures for men and women.
3. Measure blood pressure with a sphygmomanometer.
4. Demonstrate the effects of posture on blood pressure.
5. Demonstrate the effect of cold on blood pressure.
6. Calculate venous pressure.
7. Perform the Schneider, Harvard Step, and Tuttle Pulse-Ratio tests.

Materials

Wooden step model
Stethoscopes
Sphygmomanometers
Alcohol swabs
Meter sticks
Beakers (500 ml)
Ice bucket with ice

Introduction

Arterial blood pressure is the driving force that maintains blood flow through the capillary beds of the body. The pressure that blood exerts against the walls of blood vessels is determined by the *cardiac output* and the *peripheral resistance*. Cardiac output depends on heart rate and stroke volume, whereas peripheral resistance depends on blood viscosity, vessel length, and vessel radius. As blood flows through a vessel, the blood nearest the wall of the vessel moves more slowly than blood in the center of the vessel. It is this frictional resistance to flow that the heart must overcome to maintain adequate circulation.

Systolic blood pressure is a measure of the force exerted by the ventricles of the heart. *Diastolic blood pressure* is a measure of the force exerted by the blood upon arterial walls when the ventricles are relaxed. Diastolic blood pressure is maintained by the elastic recoil of arteries and hence is a measure of peripheral resistance. Blood pressure is usually given as the ratio of systolic to diastolic pressure (120/80 mm Hg for men; 110/70 mm Hg for women).

It is better to speak of a normal *range* of blood pressures than a normal blood pressure per se. There is considerable natural variation in blood pressure from one person to another and within the same person under different conditions. In general, both systolic and diastolic pressures increase with age, physical size, and exercise.

The *pulse pressure* is the numerical difference between the systolic and diastolic pressures. The pulse pressure is a measure of the extent and efficiency of blood flow, and is normally about 40 mm Hg. A depressed pulse pressure indicates that the blood is not under sufficient pressure to effectively circulate through the capillary beds of the body. The *pulse* is a mechanical pressure wave initiated by the ejection of blood from the left ventricle. The pulse is not propagated through the blood, but it is propagated as a transient expansion of the arterial walls. The pulse rate and the heart rate are identical, but the pulse velocity is 4 to 5 m/sec—about ten times faster than blood velocity.

Experimentally, blood pressure may be measured *directly* by insertion of a catheter into an artery and measurement of the pressure with a mercury manometer or a pressure transducer. From such experiments, the actual pressures inside the chambers of the heart and the major blood vessels have been measured. However, blood pressure is usually measured *indirectly* by using a sphygmomanometer and a stethoscope. The sphygmomanometer cuff is used to apply pressure to an artery, and the stethoscope is used to listen to arterial sounds. Indirect blood pressure measurement provides information about the pumping efficiency of the heart (systolic BP) and about the condition of the blood vessels (diastolic BP).

Experimental Procedure

11.1. <u>Effect of Posture on Blood Pressure</u>. Have your partner seat himself on a stool and place his right arm on the lab bench so that his arm is approximately at heart level. If he has a long-sleeved shirt on, the sleeve should be rolled up as far as possible. Wrap a deflated sphygmomanometer cuff around his upper arm about 1 inch above the antecubital space, as shown in Figure 11.1. Line up the arrow on the cuff with the brachial artery and fasten the cuff snugly with the Velcro strip or hook.

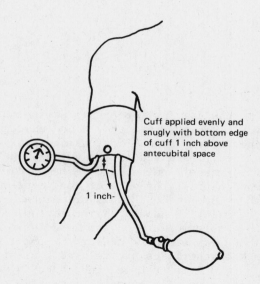

Cuff applied evenly and snugly with bottom edge of cuff 1 inch above antecubital space

1 inch-

Figure 11.1 Positioning of the sphygmomanometer cuff.

130

Find the pulse in the brachial artery just above the bend of the elbow on the medial margin of the biceps brachii muscle. Clean the earpieces of a stethoscope with an alcohol swab before putting the earpieces in your ears. Then place the stethoscope bell firmly over the brachial artery where you have felt the pulse. Hold the rubber bulb of the sphygmomanometer in the palm of your hand with the thumbscrew between your thumb and index fingers. Close the thumbscrew and squeeze the bulb with the palm of your hand and the other three fingers. Continue inflating the cuff until the air pressure in the cuff exceeds the blood pressure in the brachial artery (about 180 mm Hg). Get ready to listen for systolic and diastolic sounds with the stethoscope.

Slowly release the air from the cuff by opening the thumbscrew. Take the systolic blood pressure reading when you hear a loud, rapping sound. This sound occurs when blood is forced through the partially occluded brachial artery by ventricular systole. The turbulent arterial blood flow and the sudden expansion of the arterial wall produce the rapping or *Korotkoff sounds*. As the air pressure is reduced further, the arterial blood flow becomes less turbulent, and the sounds become louder, then muffled, and finally disappear. Take the diastolic blood pressure reading when you hear the sounds first become muffled. Wait 2 minutes and repeat the procedure.

1. What is your partner's mean systolic blood pressure when sitting?

2. What is your partner's mean diastolic blood pressure when sitting?

3. What is your partner's mean pulse pressure when sitting?

Have your partner recline on the lab bench for 5 minutes. Determine his blood pressure again.

4. What is your partner's blood pressure when reclining?

5. What is your partner's pulse pressure when reclining?

Have your partner stand up. It is important that when he stands up, he stands *still*. Take his blood pressure immediately upon standing. If the Korotkoff sounds are indistinct when he is standing, have your partner raise his arm over his head to empty his veins. Then inflate the cuff and have him lower his arm.

6. What is your partner's blood pressure when standing?

<u>7</u>. What is your partner's pulse pressure when standing?

<u>11.2</u>. <u>Effect of Cold on Blood Pressure</u>. Have your partner seat herself on the lab bench. Determine her blood pressure as described in the previous section. Then have her immerse her hand in a beaker of ice water for 15 seconds and then remove it. Redetermine her blood pressure every 20 seconds for 2 minutes.

<u>8</u>. Does the blood pressure increase, decrease or remain the same over the 2-minute period?

<u>11.3</u>. <u>Venous Pressure Determination</u>. Have your partner seat herself at the lab bench. To determine venous pressure, have her hold her hand at the level of her heart and then slowly raise her hand until the veins in the hand collapse. Measure the height to which the hand was raised above heart level by using a meter stick. Obtain your reading in centimeters; this is the venous pressure in centimeters of water. To obtain the venous pressure in millimeters of mercury, substitute this value in the following equation: mm Hg = cm(10)/13.6.

<u>9</u>. What is your partner's venous pressure?

<u>11.4</u>. <u>The Schneider Test</u>. The *Schneider test* is a test of circulatory efficiency, that is, it is a test of cardiovascular fitness rather than a measure of athletic condition. For this test, systolic blood pressure and pulse rate will be measured. Use the procedure described earlier for the blood pressure measurements. During the test, it is important that your partner exercise in a predefined manner by using a step model. The step model should have two steps with a combined height of 20 inches for males and 16 inches for females. When performing the step exercise, use both steps of the model and step up on the model with the left foot and then go to the top step with the right foot. Step down from the model by reversing your steps.

Have your partner recline for 5 minutes. Attach a sphygmomanometer cuff and obtain a reclining *systolic* blood pressure reading. Obtain a reclining pulse rate by counting the pulse felt in the radial artery of the wrist for 15 seconds and multiplying by four to get the pulse rate per minute. Enter these values in the table provided on the data sheet. Then have your partner stand up for 2 minutes. Take a new systolic blood pressure reading and a new pulse rate immediately upon standing. Then have your partner exercise on the step model by stepping up and down on the model once every 3 seconds for a total time of 15 seconds. At the end of the exercise, your partner should stand at ease, and the pulse rate should be taken immediately for 15 seconds. The next 15-second interval is skipped and the pulse rate is then taken again for 15 seconds, and so on, for a period of 3 minutes. Each 15-second pulse rate is then multiplied by four to get pulse rate per minute. The score achieved on the Schneider test is calculated by using the charts shown in Table 11.1. The total score is found by adding the scores obtained in each section. A perfect score is 18, and a score less than 9 indicates poor cardiovascular fitness.

<u>11.5</u>. <u>The Harvard Step Test</u>. The *Harvard Step test* measures general endurance but does not give specific information about physical strength, muscular endurance,

132

Table 11.1 Schneider Scoring Table

RECLINING PULSE RATE

		PULSE RATE INCREASE ON STANDING (Points)				
Rate	Points	0-10 beats	11-18 beats	19-26 beats	27-34 beats	35-42 beats
50-60	3	3	3	2	1	0
61-70	3	3	2	1	0	-1
71-80	2	3	2	0	-1	-2
81-90	1	2	1	-1	-2	-3
91-100	0	1	0	-2	-3	-3
101-110	-1	0	-1	-3	-3	-3

STANDING PULSE RATE

		PULSE RATE INCREASE IMMEDIATELY AFTER EXERCISE (Points)				
Rate	Points	0-10 beats	11-20 beats	21-30 beats	31-40 beats	41-50 beats
60-70	3	3	3	2	1	0
71-80	3	3	2	1	0	0
81-90	2	3	2	1	0	-1
91-100	1	2	1	0	-1	-2
101-110	1	1	0	-1	-2	-3
111-120	0	1	-1	-2	-3	-3
121-130	0	0	-2	-3	-3	-3
131-140	-1	0	-3	-3	-3	-3

RETURN OF PULSE RATE TO STANDING NORMAL AFTER EXERCISE

Seconds	Points
0-30	3
31-60	2
61-90	1
91-120	0
Above 120	-1

SYSTOLIC PRESSURE STANDING COMPARED WITH RECLINING

Change in mm Hg	Points
Rise of 8 or more	3
Rise of 2-7	2
No rise	1
Fall of 2-5	0
Fall of 6 or more	-1

133

or cardiovascular endurance. The test measures endurance by evaluating the pulse response to exercising on a 20-inch step.

Have your partner step up and down on the 20-inch step model at the rate of 30 complete steps per minute for as long as possible to a maximum of 5 minutes. Since this is a test of endurance, it is important to try to last for the full 5-minute period. However, if your partner is unable to continue stepping at the proper cadence, stop the exercise and record the length of time spent exercising in *seconds*. As soon as the exercise is over, have your partner sit down at the lab bench. Your partner's pulse should be counted for the following time periods: 1 to 1½ minutes after exercise, 2 to 2½ minutes after exercise, 3 to 3½ minutes after exercise. These three pulse counts should then be added together and substituted in the following equation:

$$\text{Score} = \frac{\text{Duration of exercise (sec) X 100}}{2 \text{ X sum of pulse rates during recovery}}$$

The interpretation of scores is as follows:

Below 55—poor physical condition
55 to 64—low average condition
65 to 79—high average condition
80 to 90—above average condition
Above 90—excellent condition

11.6. The Tuttle Pulse-Ratio Test. The *Tuttle Pulse-Ratio test* measures physical condition and is useful in detecting heart abnormalities. The test is based on the observation that a "less fit" person will have a higher pulse rate during recovery from exercise than a "more fit" person. The cardiovascular efficiency of a person is determined by the amount of exercise required to obtain a pulse ratio of 2.5.

Have your partner seat himself at the lab bench and obtain a resting pulse rate for 15 seconds. Multiply this value by four to get the resting pulse rate per minute. In this test, male subjects should take 20 steps and female subjects should take 15 steps on the model. Record the number of steps that your partner takes during the test on the data sheet (S_1). Immediately after the stepping exercise, have your partner sit down again and count the pulse for 2 minutes. This total postexercise rate is divided by the resting rate to give the first pulse ratio (r_1). After he has sat until his pulse rate has returned to normal, have your partner exercise on the step model again. This time, male subjects should take 40 complete steps and female subjects should take 35 complete steps in 1 minute on the step model (S_2). Immediately after the exercise, have your partner sit down and again count his pulse for 2 minutes. The second pulse ratio is obtained by dividing the resting rate into the total second postexercise rate.

The number of steps required to obtain a pulse-ratio of 2.5 is calculated from the following equation:

$$S_0 = S_1 + \frac{(S_2 - S_1) \text{ X } (2.5 - r_1)}{r_2 - r_1}$$

In this equation,

S_0 = number of steps required to obtain a 2.5 ratio
S_1 = number of steps in the first exercise
S_2 = number of steps in the second exercise
r_1 = pulse-ratio for the first exercise
r_2 = pulse-ratio for the second exercise

The interpretation of scores is as follows:

	S_0
Boys (10 to 12 years)	33 steps
Boys (13 to 18 years)	30 steps
Adult men	29 steps
Adult women	25 steps

DATA SHEET FOR EXPERIMENT 11: EXERCISE PHYSIOLOGY

Name _____ Lab Section _____ Date _____

11.1. Effect of Posture on Blood Pressure

1. What is your partner's mean systolic blood pressure when sitting?

_____

2. What is your partner's mean diastolic blood pressure when sitting?

_____

3. What is your partner's mean pulse pressure when sitting?

_____

4. What is your partner's blood pressure when reclining?

_____

5. What is your partner's pulse pressure when reclining?

_____

6. What is the blood pressure when standing?

_____

7. What is the pulse pressure when standing?

_____

11.2. Effect of Cold on Blood Pressure

Initial blood pressure _____

Final blood pressure _____

8. Does the blood pressure increase, decrease, or remain the same over the 2-minute

period? _____

11.3. Venous Pressure Determination

9. What is your partner's venous pressure? _____

137

11.4. The Schneider Test

Reclining systolic blood pressure _____

Reclining pulse rate _____

Standing systolic blood pressure _____

Standing pulse rate _____

Pulse rate—0 to 15 seconds _____

Pulse rate—30 to 45 seconds _____

Pulse rate—60 to 75 seconds _____

Pulse rate—90 to 110 seconds _____

Condition	Score
Reclining pulse rate	
Pulse rate increase on standing	
Standing pulse rate	
Pulse rate increase after exercise	
Return of pulse rate to standing normal	
Systolic pressure standing compared to reclining	
Total Score	

11.5. The Harvard Step Test

Duration of exercise (seconds) _____

Pulse rate—1 to $1\frac{1}{2}$ minutes _____

Pulse rate—2 to $2\frac{1}{2}$ minutes _____

Pulse rate—3 to $3\frac{1}{2}$ minutes _____

Sum of pulse rates _____

Score _____

11.6. The Tuttle Pulse-Ratio Test

Resting pulse rate   _____

S_1 = number of steps in first exercise   _____

S_2 = number of steps in second exercise   _____

r_1 = pulse-ratio for first exercise   _____

r_2 = pulse-ratio for second exercise   _____

S_0 = number of steps for 2.5 ratio   _____

URINALYSIS

Behavioral Objectives

The student should be able to:

1. Describe the Fick method for determining renal blood flow.
2. State the relationship among clearance, urine volume, and urine concentration.
3. Calculate glomerular filtration rate from sample data.
4. Identify the normal physical characteristics of urine.
5. Perform standard tests for urine glucose, albumin, ketones, and so on.
6. Solve problems involving clearance and GFR equations.

Materials

Disposable urine collection cups
Disposable Pasteur pipettes with dropper bulbs
Wax marking pencils
Urinometers with reservoirs
Graduated cylinders (100 ml)
Graduated cylinders (10 ml)
Beakers (50 ml)
Small glass funnels
Beakers (250 ml)
Large test tubes
Metal test tube racks
Hot plates
pH paper (4 to 8 range)
Filter paper (small size)
Benedict's solution
Rothera's reagent
Ammonium hydroxide (conc.)
Barium chloride (10 percent)
Silver nitrate (10 percent)
Harrison's (Fouchet's) reagent
Sulkowitch's reagent
Hydrochloric acid (0.1 N)
Sulfosalicylic acid (3 percent)

Introduction

In the kidneys, a fluid that resembles blood plasma is produced by glomerular filtration. As this filtrate passes through the kidney tubules, its volume is decreased and its composition is altered by tubular reabsorption and secretion to form *urine*. In this process, wastes are eliminated while water, important electrolytes, and metabolites are conserved.

In a resting adult, the kidneys receive about 25 percent of the cardiac output. Renal blood flow may be measured by using a flowmeter, or it may be calculated by applying the *Fick principle*. In the Fick method, the amount of a particular substance which is taken up per unit time is measured and is then divided by the arterial-venous concentration difference across the kidney for the substance. The substance used must not be metabolized, stored, or synthesized by the kidney, and it must not affect renal blood flow. The substances of choice are *para-aminohippuric acid* (PAH) and iodopyracet. These substances are filtered and secreted such that 90 percent of their molecules are removed from the blood in a single pass through the kidneys. As a result, renal plasma flow has been calculated by dividing the amount of PAH in the urine by the plasma PAH concentration and ignoring the PAH concentration in the venous blood. The value obtained is the *"effective" renal plasma flow* (ERPF) and averages about 625 ml/min in humans.

The ERPF determined as outlined above is defined as the *clearance* of a substance, as illustrated in the example below.

Conc. of PAH in urine (U_{PAH}) $=$ 15 mg/ml

Urine volume (V) $=$ 1 ml/min

Conc. of PAH in plasma (P_{PAH}) $=$ 0.02 mg/ml

$$\text{ERPF} = \frac{U_{PAH}V}{P_{PAH}} = \frac{15(1)}{0.02} = 750 \text{ ml/min}$$

Clearance can be converted to *"actual" renal plasma flow* (RPF) by dividing by the extraction ratio:

Arterial concentration – venous concentration
Arterial concentration

If it is assumed that the extraction ratio for PAH is 0.9, then

$$\text{RPF} = \frac{\text{ERPF}}{\text{Extraction ratio}} = \frac{750}{0.9} = 833 \text{ ml/min}$$

From the RPF, the renal blood flow can be calculated by dividing by one minus the hematocrit. If it is assumed that the hematocrit is 45 percent, then

$$\text{Renal blood flow} = \frac{\text{RPF}}{1 - \text{Hct}} = \frac{833}{0.55} = 1514 \text{ ml/min}$$

The *glomerular filtration rate* (GFR) can be measured by measuring the excretion and plasma concentration of a substance that is filtered but is not reabsorbed nor secreted. The GFR is thus identical to the clearance as defined above. A substance suitable for measuring the GFR must not be metabolized, stored or synthesized in the kidney, protein-bound, toxic, or affect filtration rate. The substance of choice is *inulin*, a polymer of fructose isolated from dahlia tubers. To measure GFR, a loading dose of inulin is administered intravenously followed by a sustained infusion to maintain the arterial concentration of inulin constant. After

142

equilibration of inulin with the body fluids, a urine sample and a plasma sample are collected. Urine and plasma inulin concentrations are then determined and the GFR can be calculated. For example,

$$U_{In} = 30 \text{ mg/ml}$$

$$V = 1 \text{ ml/min}$$

$$P_{In} = 0.30 \text{ mg/min}$$

$$GFR_{In} = \frac{U_{In}V}{P_{In}} = \frac{30(1)}{0.30} = 100 \text{ ml/min}$$

In some experimental animals, the clearance of *creatinine* can also be used to measure the GFR. However, in humans, some creatinine is reabsorbed and secreted. In addition, plasma creatinine determinations tend to be inaccurate at low creatinine levels. In spite of these shortcomings, the clearance of endogenous creatinine is frequently measured. The values obtained agree fairly well with GFR values obtained using inulin because the errors in measurement with both substances tend to cancel out. Thus endogenous creatinine clearance is a useful index of kidney function, but when precise GFR measurements are needed, an inulin clearance measurement should be made.

Experimental Procedure

The best sample for a routine urinalysis is a urine specimen collected the first thing in the morning. In addition, a "midstream catch" should be used. This is done by filling the collection cup with urine after the urination has begun for a few seconds.

Obtain a urine sample as described above in the disposable urine collection cup provided. One urine sample should be sufficient for each pair of students.

12.1. Color. The normal color of urine is amber due to pigments that result from bile metabolism. Various drugs (for example, sulfa drugs) and dietary substances (for example, beets) may also produce a colored urine. The color of urine changes in many disease states because of pigments that are not normally present. In addition, the color of urine darkens on standing due to the oxidation of urobilinogen to urobilin.

1. What is the color of your urine sample? Is it pale yellow, amber, reddish, brown, or black?

12.2. Character. The character or turbidity of urine refers to its ability to transmit light. Normal urine may be transparent or turbid. Turbidity may result from the precipitation of phosphates (in alkaline urine), urates (in acidic urine), or from urinary tract infections. The transparency of urine usually decreases on standing or refrigeration.

2. What is the character of your urine sample? Is it clear, slightly cloudy, cloudy, turbid, very turbid, or milky?

143

12.3. **Odor.** The odor of urine is normally aromatic. This odor is thought to be the result of volatile acids. Urine that has been standing for any length of time develops a foul ammonialike odor due to the decomposition of urea. Although urine may have a characteristic odor in various disease states (for example, sweet smell in diabetes mellitus), the odor of urine is not considered especially significant and is not reported in laboratory tests.

3. **What is the odor of your urine sample? Is it aromatic, ammonialike, or sweet?**

12.4. **Specific Gravity.** The normal specific gravity of urine varies between 1.005 and 1.030. The specific gravity is a measure of urine concentration, a specific gravity of 1.005 indicating a dilute urine and a specific gravity of 1.030 indicating a concentrated urine. Sodium chloride and urea are the two principal solids found in urine, and hence they have the greatest effect on specific gravity. The NaCl concentration reflects the amount of salt in the diet; the urea concentration reflects the amount of protein in the diet.

Measure the specific gravity of your urine sample by pouring some urine into the reservoir of a urinometer. Gently place the urinometer float in the reservoir and spin it like a top. If the urinometer doesn't float, add more urine to the reservoir. Make sure that the float is not sticking to the side of the reservoir. The specific gravity is measured by reading the urinometer scale at the urine-air interface.

4. **What is the specific gravity of your urine sample?**

12.5. **pH.** The normal pH of urine is between 5.0 and 8.5, with a mean pH of 6.0. The measurement of urine pH is used in the determination of acidosis and alkalosis. In addition, urine pH is used to monitor the effectiveness of specific medication programs designed to produce an acidic or alkaline urine.

Obtain a strip of pH paper. This strip of paper has been impregnated with indicators that change color in the presence of different concentrations of hydrogen ions. Dip the pH paper into your urine sample and compare its color with the colors on the package that comes with the paper.

5. **What is the pH of your urine sample?**

12.6. **Albumin.** Albumin is a plasma protein that is normally absent from urine since it is too large to be filtered. The presence of albumin (albuminuria) is a sign of kidney disease or may occur normally after strenuous exercise.

Add 1 ml of urine to a test tube. Then add 3 ml of 3 percent sulfosalicylic acid to the test tube. If the mixture becomes turbid or forms a precipitate, albumin or some other plasma protein is present.

6. **Does your sample show no reaction, a faint trace, a trace, or a moderate amount of albumin?**

12.7. Glucose. Glucose is not normally found in urine. However, it may be found in the urine after strenuous exercise, ingestion of large amounts of glucose, emotional disturbances, or in diabetes mellitus.

Place 10 drops of urine in a test tube. Add 5 ml of Benedict's solution and place the test tube in a boiling water bath for 5 minutes. Then remove the test tube by using a test tube holder and note any color change that has occurred. In Benedict's test, glucose molecules reduce the copper of Benedict's reagent to form cuprous oxide, a colored compound. As the amount of cuprous oxide increases, the color of the solution varies from blue to green, yellow, or orange.

<u>7.</u>   <u>Does your urine sample contain glucose</u>?

Indicate the color that your sample turned.

12.8. Ketones. Ketones are not normally found in urine. If the body is unable to oxidize glucose, as in diabetes mellitus or low carbohydrate diets, fats are oxidized instead. In this fat catabolism, acetoacetic acid, β-hydroxybutyric acid, and acetone are produced. The test for ketones involves testing for these breakdown products of fat metabolism.

Place 5 ml of urine in a test tube and add 1 ml of Rothera's reagent. Tilt the tube and carefully add 1 ml of concentrated ammonium hydroxide by allowing it to gently flow down the sides of the tube. If ketones are present, a pink-purple ring of ammonium sulfate will develop at the urine-Rothera's reagent interface. A negative test shows no ring or a brown ring.

<u>8.</u>   <u>Does your urine sample contain ketones</u>?

12.9. Bile. Bile is not normally found in urine. It is present in urine in infectious hepatitis, obstructive jaundice, and in various liver diseases.

Place 10 ml of urine in a test tube. Add 1 ml of 0.1-N HCl to the test tube. Then add 5 ml of 10-percent barium chloride. Shake the tube and filter the mixture through a piece of filter paper into a beaker. Add 1 drop of Harrison's (Fouchet's) reagent to the residue on the filter paper. If bile is present, a blue-green color will result.

<u>9.</u>   <u>Does your urine sample contain bile</u>?

12.10. Calcium. Calcium, in contrast to the substances tested above, is normally found in urine. Increased amounts of calcium are present in hyperparathyroidism and in bone tumors; decreased amounts of calcium are present in hypoparathyroidism.

Place 2 ml of urine in a test tube. Add 5 ml of Sulkowitch's reagent to the test tube. Mix the contents of the tube and allow it to stand for 3 minutes. If a fine white cloud appears, the calcium content is normal. If no white cloud appears, the calcium content is below normal; if a heavy white precipitate forms, the calcium content is elevated.

10. Does your urine sample have normal, decreased, or elevated amounts of calcium?

12.11. Chlorides. Chlorides are normally found in urine and are used as an indicator of dietary salt intake. People on a restricted salt diet will usually excrete less than 0.6 percent NaCl in the urine. Most methods used for the determination of chlorides are based on the precipitation of chlorides as insoluble salts.

Place 1 ml of urine in a test tube. Add 1 drop of silver nitrate solution. If a precipitate forms, chlorides are present.

11. Does your urine sample contain chlorides? If so, what is the precipitate formed?

12.12. Sulfates. Sulfates are normally found in the urine. Most methods used for the determination of sulfates are based on the precipitation of sulfates as insoluble salts.

Place 1 ml of urine in a test tube. Add 1 drop of barium chloride. If a precipitate forms, sulfates are present.

12. Does your urine sample contain sulfates? If so, what is the precipitate formed?

12.13. Problems.

1. If inulin appears in the urine at the rate of 25 mg/hr, and its plasma concentration is 1 mg/liter, what is the glomerular filtration rate?

2. If the volume of urine produced in 1 hr was 150 ml, how much was the filtrate concentrated by reabsorption of water in problem 1 above?

3. If the plasma inulin concentration is maintained at 0.1 mg/liter and 2 mg of inulin appear in the urine, what is the glomerular filtration rate?

4. Using the inulin method, a GFR of 8000 ml/hr was determined. The amount of urine entering the bladder during this time period was 50 ml. How much did the concentration of inulin increase as it passed from Bowman's capsule to the bladder?

5. Inulin and urea were each maintained at 2 mg/ml in the plasma. In 1 minute, the following appeared in the urine:

Inulin   250 mg
Urea   100 mg
Water   2 ml

What is the glomerular filtration rate? How fast are urea and water reabsorbed from the tubules?

6. Under the same conditions as in problem 5 above, the following appeared in the urine in 1 hr:

Inulin   12 g
Urea   5 g
Water   100 ml

What is the glomerular filtration rate? At what rate are urea and water being reabsorbed from the tubules?

DATA SHEET FOR EXPERIMENT 12: URINALYSIS

Name_____ Lab Section _____ Date _____

12.1. Color

1. What is the color of your urine sample? _____

12.2. Character

2. What is the character of your urine sample? _____

12.3. Odor

3. What is the odor of your urine sample? _____

12.4. Specific Gravity

4. What is the specific gravity of your urine sample?

_____

12.5. pH

5. What is the pH of your urine sample? _____

12.6. Albumin

6. Does your sample show no reaction, a faint trace, a trace, or a moderate

amount of albumin? _____

12.7. Glucose

7. Does your urine sample contain glucose? _____

Color after Benedict's test _____

12.8. Ketones

8. Does your urine sample contain ketones? _____

12.9. Bile

9. Does your urine sample contain bile? _____

12.10. Calcium

10. Does your urine sample have normal, decreased, or elevated amounts of calcium?
_____

12.11. Chlorides

11. Does your urine sample contain chlorides? _____

    If so, what is the precipitate formed? _____

12.12. Sulfates

12. Does your urine sample contain sulfates? _____

    If so, what is the precipitate formed? _____

EXPERIMENT 13

ENZYMATIC DIGESTION

Behavioral Objectives

The student should be able to:

1. List the end products of chemical digestion.
2. Demonstrate the effect of temperature on digestion.
3. Demonstrate the effect of rennin on protein.
4. Demonstrate the effect of amylase on starch.
5. Perform a Benedict's test for reducing sugars.
6. Demonstrate the effect of lipase on fat.

Materials

Heavy cream and skim milk
Litmus indicator powder
Wax marking pencils
Hot water bath (37°C)
Ice bath
Large test tubes
Beakers (250 ml)
Beakers (50 ml)
Graduated cylinders (10 ml)
Stirring rods
Test tube racks
Hotplates
Rennin powder
Lipase powder
Spatulas
Porcelain spot plates
Pasteur pipettes with dropper bulbs
Parafilm
Benedict's solution
Boiled starch (1 percent)
Iodine potassium iodide
Maltose (1 percent)

151

Introduction

Digestive enzymes accelerate the hydrolysis of carbohydrates, proteins, and lipids into simple sugars, amino acids, and fatty acids and glycerol, respectively. *Hydrolysis* is a chemical reaction in which large molecules are split into smaller molecules by combining with water molecules. Digestive enzymes are complex protein molecules which accelerate the rate of these hydrolysis reactions without being altered by the reaction themselves. Complex protein molecules have a three-dimensional structure which depends on the sequence of amino acids that make up the protein. A protein molecule which functions as an enzyme is thought to have an active site where it can bind to a substrate molecule to form an enzyme-substrate complex. If two substrates or reactants are bound by an enzyme (for example, a carbohydrate molecule and a water molecule), the enzyme provides a surface upon which a reaction can occur. Since the enzyme is an innocent bystander in this process, the enzyme molecule will be unchanged by the reaction that occurs between the substrate molecules. The binding of the substrates to the enzyme lowers the energy required for the reaction to occur, and, as a result, speeds up the reaction rate.

The rate of a chemical reaction depends on several factors: enzyme concentration, substrate concentration, and product concentration. A detailed analysis of reaction rates (kinetics) is beyond the scope of this laboratory manual. However, the rate-limiting step in enzyme-catalyzed reactions is usually the rate of formation of the enzyme-substrate complex. Any factor which alters the ability of the enzyme to bind substrate will consequently alter reaction rate. All enzymes in the body have certain pH and temperature optima. The rates of enzymatic reactions can therefore be altered by shifting either the pH or the temperature away from these optimal conditions. For example, if the temperature is lowered, the kinetic energy of the enzyme and substrate molecules in solution will be lowered, and it will be more difficult for the enzyme-substrate complex to form. In addition, the three-dimensional shape of the active site may be changed so as to inhibit binding of the substrates. Most enzymatic reactions in the body have a temperature optimum of 37°C (normal body temperature) and a pH optimum of 7.4 (normal body pH). However, digestive enzymes frequently have pH optima much higher or lower than 7.4 to coincide with their functions in the stomach or small intestine.

Experimental Procedure

13.1. **Effect of Temperature on Lipase.** Lipase is an enzyme that hydrolyzes fats into fatty acids and glycerol. This part of the experiment will demonstrate the effect of lipase on the fats present in heavy cream.

Obtain two test tubes and label them A and B with a wax marking pencil. In addition, place your initials on each of the test tubes. Place 5 ml of litmus cream into each test tube. Litmus cream is made by mixing litmus indicator powder with heavy cream in a beaker. After placing the litmus cream in the two test tubes, add a small amount of lipase powder to each tube with a spatula and stir the tubes with a stirring rod. Place tube A in an ice bath and tube B in a 37°C hot water bath. Allow the tubes to equilibrate for 1 hour.

1. What color are the tubes before equilibration?

2. What color is tube A after equilibration?

3. What color is tube B after equilibration?

4. Has a reaction occurred in each tube? Why or why not?

13.2. Effect of Rennin on Protein. Rennin is an enzyme that coagulates proteins. This part of the experiment will demonstrate the effect of rennin on the
proteins present in skim milk.
    Obtain two test tubes and label them A and B with a wax marking pencil. Place
your initials on each tube as well. Add 5 ml of skim milk to each tube. Then add
a small amount of rennin powder to tube A only. Stir both tubes with a stirring
rod. Place both tubes in the 37°C water bath and allow them to equilibrate for
1 hour.

5. Describe the contents of each tube before equilibration.

6. Describe the contents of tube A after equilibration.

7. Describe the contents of tube B after equilibration.

8. Has a reaction occurred in each tube? Why or why not?

13.3. Effect of Amylase on Starch. Salivary amylase is an enzyme that hydrolyzes starch into soluble starch, dextrin, archrodextrin, erythrodextrin, and
maltose. This part of the experiment will demonstrate the effect of salivary
amylase on the hydrolysis of starch.
    Place three drops of boiled starch in a depression in a porcelain spot plate.
Add a drop of IKI to the starch.

9. What color results?

    Place 3 drops of maltose in a different depression of the spot plate. Add a
drop of IKI to the maltose.

10. What color results?

153

Set up a boiling water bath by filling a 250-ml beaker half full of water and placing it on a hotplate. Obtain two test tubes and label them A and B, along with your initials. Add 5 ml of maltose and 5 ml of Benedict's solution to tube A. Add 5 ml of boiled starch and 5 ml of Benedict's solution to tube B. Stir both tubes with a stirring rod and place them in the boiling water bath for 5 minutes.

11. <u>What color does tube A turn?</u>

12. <u>What color does tube B turn?</u>

The colors signify the following sugar concentrations:

Blue—no sugar present
Green—trace of sugar
Yellow—low concentration of sugar
Orange—high concentration of sugar
Red—very high concentration of sugar

The above colors result from the reaction of cupric ions in the Benedict's solution with reducing sugars like maltose.

The spot plate method and Benedict's test described above will serve as controls for the procedure outlined below. They provide two qualitative methods for determining the presence of substrate (starch) and the breakdown products of starch digestion (maltose).

Collect 5 ml of saliva (not bubbles) in a 50-ml beaker. Chew on a small piece of parafilm to stimulate salivation. It is important that you do not use gum to stimulate salivation because gum contains sweeteners that will affect the results. Add 10 ml of boiled starch to the saliva in the beaker and mix the solutions with a stirring rod. Then immediately test the mixture for the presence of starch by the spot plate method described above. Continue testing 3-drop portions of the mixture until starch is no longer present. Use time intervals of 30 seconds between tests.

13. <u>How long does it take for complete hydrolysis?</u>

Then test a 1-ml sample of the mixture for maltose by using a Benedict's test.

14. <u>What is the color of the tube after 5 minutes?</u>

15. <u>How much sugar is present?</u>

DATA SHEET FOR EXPERIMENT 13: ENZYMATIC DIGESTION

Name _____ Lab Section _____ Date _____

13.1. Effect of Temperature on Lipase

1. What color are the tubes before equilibration? _____

2. What color is tube A after equilibration? _____

3. What color is tube B after equilibration? _____

4. Has a reaction occurred in each tube? _____

   Why or why not? _____

   _____

13.2. Effect of Rennin on Protein

5. Describe the contents of each tube before equilibration.

   _____

   _____

6. Describe the contents of tube A after equilibration.

   _____

   _____

7. Describe the contents of tube B after equilibration.

   _____

   _____

8. Has a reaction occurred in each tube? _____

   Why or why not? _____

13.3. Effect of Amylase on Starch

9. What color results? _____

10. What color results? _____

11. What color does tube A turn? _____

12. What color does tube B turn? _____

13. How long does it take for complete hydrolysis? _____

155

14. What is the color of the tube after 5 minutes? _____

15. How much sugar is present?_____

COMPUTERS IN PHYSIOLOGY

Behavioral Objectives

The student should be able to:

1. Give examples of physiological applications of computers.
2. Describe the usefulness of the analog computer in physiology.
3. Describe the usefulness of the digital computer in physiology.
4. State the parts of a computer system and describe how they are interrelated.
5. Define machine language, assembly language, and higher level languages.
6. Define statements, variables, literals, and algorithms.
7. State the general application of each of the common computer languages.
8. Enter and execute a BASIC computer program.

Materials

Computer with BASIC language processor
Computer terminal

Introduction

14.1. Computers and Physiology. Computers are used in almost every scientific field. A *computer* is a device that is used to process information (data), and a wide variety of different types of information lend themselves to computer analysis. The simplest uses of a computer are to add, subtract, multiply, and divide. More complex calculations can involve the solution of simultaneous equations, differential equations, integration, and so on. In physiology, since natural variation must be considered, computers have been used to perform statistical analyses. Computers have been used to record several responses to similar stimuli and to average the responses to see if a trend is present. Computers have been used to perform time-consuming, repetitive calculations that would not be practical to perform by hand. Computers have been used to perform transformations on raw data and to plot the results graphically or in tabular form. The data can then be stored and retrieved for future reference. Thus a computer is useful in physiology and other disciplines because it can process large quantities of information much more rapidly than an individual or a group of individuals.

14.2. The Analog Computer. There are two basic types of modern electronic computers: analog and digital. *Analog computers* work with voltages or currents which stand for physiological variables (blood pressure, heart rate, force of

157

contraction, action potential frequency, and so on). The power of an analog computer lies in its ability to solve several complicated equations simultaneously. Thus analog computers are useful in the simulation, modeling, and prediction of physiological behavior. An analog computer is programmed to solve a particular equation or set of equations by making connections with patchcords on a circuit board. Changing from one equation to another often involves repatching the entire circuit. Since it is often impossible to write an exact equation to describe a physiological phenomenon, analog computers have limited applications in physiology.

14.3. The Digital Computer. The *digital computer* works with numbers (digits) and hence it can perform calculations with great accuracy. Digital computers can also manipulate character information (letters), which gives them an advantage over analog computers. A digital computer is programmed with a set of commands which can be fed into the computer from an input device (punched card reader, teletype, and so on). This program can be easily changed by inserting lines, deleting lines, and the like.

The digital computer must be able to recognize (read) programmed instructions, to remember the program being executed, to read input data, to perform arithmetic calculations, to read out and store results, and to control its entire operation. Thus a computer system consists of an input section, main and secondary memory sections, an arithmetic section, an output section, and a control section. The arithmetic and control sections are usually referred to as the central processing unit (CPU). Figure 14.1 summarizes the flow of information through a digital computer. The CPU is capable of understanding instructions coded as a string of 0's and 1's—a *binary code*. The control unit of the CPU carries out program instructions sequentially; the arithmetic unit of the CPU performs arithmetic and logical operations (comparisons).

The main memory and secondary memories consist of a group of *locations* which can be filled with instructions or data. Each location has a specific *address* in the memory. The secondary memories are used to extend the capacity of the main memory.

The most widely used input and output devices include cathode ray tube (CRT) screens or printers associated with a keyboard.

14.4. Computer Languages. Computer *hardware* refers to the computer itself, its circuits and its other physical parts. Computer *software* refers to the programming needed to make the computer solve a particular problem or perform a given task. As noted above, the CPU of a digital computer executes program instructions expressed in a binary code. This binary code is called *machine language*. As you can imagine, it is extremely tedious and difficult to write programs in this form. It is easier to write programs in *assembly language*, but it is still quite complicated because each type of

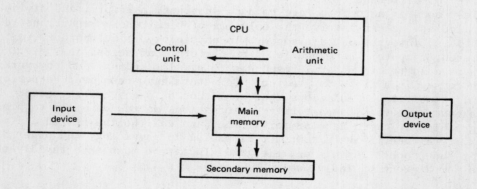

Figure 14.1 Information flow in a digital computer.

158

computer has its own assembly language instructions. As a result, most computer
programs are written in a *higher level language* which permits the problem and its
solution to be expressed in a somewhat readable form. The higher level languages
in common use are listed in Table 14.1. The computer translates the assembly
language or higher level language instructions into machine language by means of a
compiler. The sequence of events in program execution is shown in Figure 14.2.

Table 14.1. Computer Languages

LANGUAGE ACRONYM	LANGUAGE NAME	GENERAL APPLICATIONS
ALGOL	ALGOrithmic Language	Mathematical computation; Nonnumeric Manipulation
APL	A Programming Language	Arrays; Graphics; Mathematical Computation
BASIC	Beginner's All-Purpose Symbolic Instruction Code	Personal Computers; Mathematical Computation
COBOL	COmmon Business Oriented Language	File Manipulation; Nonnumeric Computation
FORTRAN	FORmula TRANSlator	Mathematical Computation
LISP	LISt Processing Language	Artificial Intelligence
PASCAL	PASCAL	Mathematical Computation; Arrays; Records
PL/1	Programming Language 1	Mathematical Computation; Nonnumeric Computation

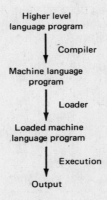

Figure 14.2 Stages in program processing.

159

BASIC is the most widely used language for personal computers. Unfortunately, different versions of BASIC have been developed for different computer systems. Nevertheless, all forms of BASIC share some common features. Any word that is used as a statement, command, function, or programming aid is called a *reserved word*. A list of reserved words for almost all versions of BASIC is given in Table 14.2.

Table 14.2 BASIC Reserved Words

AND	GOSUB	NOT	RETURN
AUTO	GOTO	ON	RUN
CLEAR	IF	OR	STOP
CONT	INPUT	OUT	TAB
DATA	LET	POKE	THEN
DIM	LIST	PRINT	TO
ELSE	NEW	READ	USR
END	NEXT	REM	WAIT
FOR			

The BASIC language has several language elements—statements, variables, literals, functions, and statement numbers. *Statements* are commands that tell the computer to perform a specific task, for example, PRINT, RUN. Every line of a BASIC program has a *statement number*, and the program will be executed in order of statement numbers. *Variables* are items whose value may change during the program. BASIC uses numeric variables to store numbers; they are represented by a letter or a letter followed by a number, for example, X, Y, Fl. BASIC uses string variables to store letters; they are represented by a letter and a dollar sign, for example, X$, R$. *Literals* are strings of characters or numbers that do not stand for any other variable or quantity, for example, "HELLO." String literals are enclosed in quotation marks and are useful for making comments; numeric literals are not enclosed in quotation marks. Functions are statements that carry out an operation, for example, SQR for square root and ABS for absolute value.

14.5. A BASIC Program. The following BASIC program will perform a paired *t* test on two groups of data where *n* = 4 for each group.

```
10 REM  THIS PROGRAM PERFORMS A PAIRED T TEST ON
20 REM  TWO GROUPS OF DATA WITH N = 4
30 PRINT "INPUT THE FIRST DATA GROUP:"
40 INPUT A1, A2, A3, A4
50 PRINT "INPUT THE SECOND DATA GROUP:"
60 INPUT B1, B2, B3, B4
70 D1 = ABS (A1 - B1)
80 D2 = ABS (A2 - B2)
90 D3 = ABS (A3 - B3)
100 D4 = ABS (A4 - B4)
110 D = D1 + D2 + D3 + D4
120 DA = D/4
130 DS = (D1*D1 + D2*D2 + D3*D3 + D4*D4) - (D*D/4)
140 S2 = DS/3
150 S = SQR (S2)
160 SD = S/SQR (4)
170 T = DA/SD
180 PRINT "T = "; T
190 END
```

Experimental Procedure

Enter the program into the memory of your computer. Your instructor will explain how to sign on to the computer and how to gain access to the input mode for creating a program. After the program has been entered, carefully check each line to insure that you have not made any typing errors. Errors can be corrected by editing the lines that you have already typed in; this process is called *debugging* and represents an important feature of computer programming. After you are sure that the program is correct, type the command RUN. Use the data from the sample problem described in Experiment 1. The numbers in each group should be separated by commas when they are entered. The value of t that the computer calculates should be 3.47.

The set of instructions (program) listed above constitutes an *algorithm* for a paired t test—a set of instructions that results in the solution of a particular problem or class of problems. To demonstrate that this algorithm can be extended to use groups of data with more than four numbers, edit the program so that it will be able to find the t value for $n = 5$.

Turn in a listing (hardcopy) of your modified program as your data sheet. Make sure that you run the program to prove that it works correctly!

Name _____ Lab Section _____ Date _____

Listing of modified t test program to use $n = 5$
Attach listing below:

Acetic acid (1 percent)
    1 ml glacial acetic acid
    Add distilled water to make 100 ml

Anti-A serum
Anti-B serum }   Commercially available from supply companies
Anti-D serum

Barium chloride
    20.81 g barium chloride
    Add distilled water to make 100 ml

Benedict's solution
    Commercially available from supply companies

Bromthymol blue
    0.4 g bromthymol blue
    500 ml of ethyl alcohol (95 percent)
    Dissolve bromthymol blue in alcohol
    Add distilled water to make 1000 ml

Cyanmethemoglobin standard solution
    Commercially available from supply companies

Drabkin's reagent
    1 g sodium bicarbonate
    0.2 g potassium ferricyanide
    0.052 g potassium cyanide
    Add distilled water to make 1000 ml

Fouchet's (Harrison's) reagent
    25 g trichloroacetic acid
    10 ml of 10 percent ferric chloride
    Add distilled water to make 100 ml

Frog Ringer (low calcium)
    6.996 g sodium chloride
    0.186 g potassium chloride
    0.121 g tris buffer
    Add distilled water to make 1000 ml
    pH to 7.25

Frog Ringer (normal)
    6.996 g sodium chloride
    0.186 g potassium chloride
    0.121 g tris buffer
    0.199 g calcium chloride
    Add distilled water to make 1000 ml
    pH to 7.25

Harrison's reagent
    See Fouchet's reagent
Hydrochloric acid (0.1 N)
    8.4 g hydrochloric acid
    Add distilled water to make 1000 ml
Iodine potassium iodide (Lugol's solution)
    1.5 g iodine
    7.5 g potassium iodide
    Add distilled water to make 500 ml
Maltose (5%)
    5 g of maltose
    Add distilled water to make 1000 ml
Quinine sulfate (0.1%)
    0.5 g quinine sulfate
    Add distilled water to make 100 ml
Rothera's reagent
    Commercially available from supply companies
Silver nitrate (10 percent)
    1.69 g silver nitrate
    Add distilled water to make 100 ml
Starch (1 percent, boiled)
    10 g cornstarch
    Add starch to boiling distilled water to make 1000 ml
Sucrose (5 percent)
    5 g sucrose
    Add distilled water to make 100 ml
Sulkowitch's reagent
    2.5 g oxalic acid
    2.5 g ammonium oxalate
    5 ml glacial acetic acid
    Add distilled water to make 150 ml

82 83 84 85 9 8 7 6 5 4 3 2 1